[illegible]s Lig[illegible]ay

Official Handbook

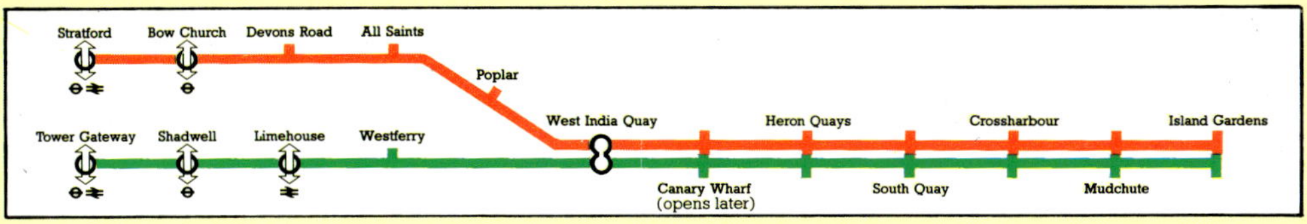

Bob Bayman and Stephen Jolly

Capital Transport

FOREWORD

About 18 months ago I wrote the foreword to the first edition of the Official Handbook and since then we have come of age with the opening to passengers on 31st August 1987.

It is appropriate to publish this second edition at this stage because we are now started on another great adventure — a tunnel under the City and a substantial series of enhancements to the Initial Railway.

So this book captures the recent past, the present and outlines the future. We hope you find it of use.

C F BONNETT
MANAGING DIRECTOR
DOCKLANDS LIGHT RAILWAY LIMITED

INTRODUCTION

"A little railway that grew very quickly" will probably be the historian's view of the Docklands Light Railway. From a 1982 paper plan to run a small Light Rail system in East London with trams, overhead wires and street running, to a 1987 automated, elevated wire-free scenic railway, to a mid-1990s system with perhaps nine times as many passengers as now and 90 vehicles and around six distinct routes! That is the summary of the growth of the DLR.

The introduction to the first book stated how the Docklands area was undergoing great change — urban renewal — on a scale unmatched in the rest of Europe: this is continuing but at an accelerated pace.

This edition finds the railway at a particularly interesting moment. Built with a cash-limited, non-inflation proofed sum of £77m through two organisations in equal amounts, the London Docklands Development Corporation (LDDC) and London Regional Transport (LRT), huge private sums of money are now helping to pay for the next phase of development — the City extension and an enhancements package to the Initial Railway.

Rather than attempt to write a beginning, middle and end text — what end should we write — we have adopted the same separate topic approach in the layout of this handbook as in the first edition. The railway defies a straight historical account because the complicated inter-weaving of politics, technology, geography, history and much else would create large breaks in the text to stop to explain other aspects.

Thanks are due to Derek Chambers, David Sexton, Joyce Porter, Brian Hardy and Mike Smith, on whose script the text is based, and for photographs to Brian Gilliam (page 14), R.J. Greenaway (page 49 upper) and R.J. Waterhouse (pages 27 upper, 43 upper and the front cover). All other photographs are by David Sexton of DLR.

ISBN 0 904711 95 1

Published by Capital Transport Publishing
38 Long Elmes, Harrow Weald, Middlesex

Printed by KPC Print, Ashford, Kent

CONTENTS

Some 80% of the route of the DLR re-uses older railway land. The descent from the junction at West India Quay to the Tower Gateway branch is built above the old docks transfer sidings and the course of the London & Blackwall Railway.

ORIGINS

Before the present railway came into existence and before the London Docklands Development Corporation was set up, London Transport had plans in the nineteen seventies to extend one of its tube lines into the Docklands area. The line, referred to as the River Line but seen always as an extension of the Fleet/Jubilee Line, would have passed under the Thames as many as five times before reaching the Thamesmead and Woolwich areas with a spur to Beckton. The great attraction of linking communities north and south of the river would have been moderated by the limited number of station sites that could have been provided. In addition, deep-level bored tube station construction below the water table is extremely expensive and it was the high cost of the scheme — £325m (close to £500m at 1988 prices) — that killed it off.

There remained a great need to improve public transport in the area, reinforced in 1981 by the creation of the Docklands Development Corporation who identified the requirement for permanent links to help attract and keep the confidence of industry and commerce. It is widely believed that the regeneration of London Docklands is the most important Inner City redevelopment work of the 1980s in the Western world. Claims that it offers a unique set of opportunities for investors and businesses to set up new enterprises are easy to believe. Whereas the up-river docks became inaccessible to larger and larger container ships, those same dock areas are strategically close to the heart of the Capital. London now has a waterscape environment covering a total of 20 sq kilometres (8 sq miles) with the remains of the magnificent dock and wharf structures within large areas of dereliction.

A Light Rail system was seen as a suitable alternative and a number of route studies were undertaken by London Transport for the Development Corporation. In June 1982 a report called 'Public Transport Provision for Docklands' was submitted to the Government. This report was jointly prepared by the Development Corporation, the Greater London Council, London Transport and the Departments of Transport, the Environment and of Industry. It recommended the building of two lines of about 12km (7½ miles), costing £65m with a west-south route and a north-south route.

In October 1982 — a remarkably short time — the Government agreed to fund the building of the Docklands Railway as a project of regional and national importance, and subsequently confirmed this under strictly controlled conditions. The outturn price was not to exceed £77m, to be funded equally by the Departments of Transport and of the Environment to the GLC and the LDDC. Both routes were to open in 1987. Although full advantage was to be taken of contemporary technology, technical solutions were not to be incorporated that could significantly increase the risk of cost-overrun or compromise the railway's completion date or reliable and economical operation. Finally, it was required that, when operational, the railway would not need revenue support from the Government.

The authorised cost figure was based on London Transport's planning estimates for a route which used a mixture of existing BR track, new and disused viaducts and cuttings, as well as a short section of street running, and which assumed a manually-driven, standard gauge, steel wheel on steel rail system with overhead current collection. Just ten stops were to be provided.

London Transport, acting as agents on behalf of the two joint clients, the GLC and the LDDC, was made responsible for preparing the Parliamentary Plans, constructing the railway and opening it on time and within budget.

From June 1984 the GLC ceased to be responsible for transport, London Transport became London Regional Transport and thus became a joint and equal client with the LDDC as well as the holder of the parliamentary powers to build the railway.

The automated version of the classic European light rail vehicle. Car 04 is seen here during pre-operational trials showing clearly the open lid of the emergency driving position. DLR uses a third rail system, the conductor rail visible to the left: most light rail systems use overhead power wires a good deal less obtrusive than the BR ones seen here near Bow.

WHAT IS LIGHT RAIL TRANSIT?

Light Rail Transit, also known as Light Rapid Transit, is a form of transportation that fits between the spheres of bus and conventional heavy railway operation. There are, it is estimated, over 300 systems in operation around the world. The whole industry has enjoyed a renaissance in the last twenty years, particularly in North America, but also in Western Europe where investment in improving the existing systems as well as opening some new ones, as in France, is taking place. In Britain, the popular Tyne & Wear Metro, serving the area in and around Newcastle and Gateshead, was the only system in operation prior to the opening of the DLR.

It can be argued that all forms of Light Rail system are hybrids because they blend features of other transit systems together in individual packages to suit individual area circumstances. The Docklands system is no exception to this rule, combining automatic operation with the retention of staff on trains, having a sophisticated signalling and control yet having no lineside signals, and keeping small and cheap stations yet having full disabled access through the provision of hydraulic lifts. A simple definition of Light Rail is difficult to achieve.

Flexibility and economy are the kingpins of Light Rail's attraction to transport operators and planners. With relatively small vehicles and the consequent possibility of tighter track curves, such systems can often weave around existing townscapes, avoiding the need for extensive urban disruption. They do not need similar levels of staffing to heavy rail, yet at the same time they can cope admirably with passenger flows far in excess of those catered for by buses. Compared with buses, Light Rail can be fast, quiet and smooth. Experience of Light Rail investment in Europe and North America over the last 30 years has shown a Light Rail system is fully able to build up traffic while at the same time relieving vehicle congestion in city centres. Additionally it offers scope for making improvements to urban environments unobtainable where conventional railway access or relatively conventional bus access is the norm.

By using lightweight overhead structures DLR avoids road intersections and the creation of isolated pockets of land, cut off by the railway. The ability of light rail to cope with gradients of up to 6% (1 in 17) and curves of 40m radius — or 25m radius on short sections — also minimises land take, reducing capital costs and disruption.

Conventional railways nearly always require gentle curves and nearly-level routes to obtain fast running speeds, and they need substantial engineering structures. Conventional railways tend to have high axle loadings on the trains and involve a substantial number of operating and engineering staff to provide a comprehensive and continuing service, but they do in return offer a fast, safe, high-capacity transport link that can serve certain well-defined corridors. Naturally, they require a lot of investment and investment levels will tend to be a lot higher than those of Light Rail. London's Underground railways are a specialised segment on their own. To move large numbers of people safely in large cities, high capacity electric trains must operate through tunnels. Construction costs are now so high that, with the possible exception of the world's most populous cities, financial justification of new metro systems is becoming impossible.

In discussing Light Rail in general, it is important not to have a narrow view of what it looks like. Both overhead and low level current collection can be used in the same system in different environmental situations, or one system or the other can be used exclusively. There is often a temptation to assume that street running — that curse of old-fashioned tram networks in Britain — is an essential part of a modern Light Rail system, but it is not. Street running does however allow further flexibility in choice of route, so that traditional railway lines, redundant railway land and street running can be combined to produce new packages of routes. Street running in the Mile End area once figured in the DLR plans, but in this particular case there were technical and environmental problems which subsequently sent the route elsewhere. Ultimately, the adoption of automatic driving for the Railway effectively ruled out street running for Docklands, yet it is still a Light Rail-based system.

Light rail can economically cater for a large range of demands, with systems around the world carrying between 2,000 and 20,000 passengers per hour. Because of the rapid development of the Docklands area, the DLR — originally designed for movements at the bottom end of the scale — will face demands somewhere near the top of the scale within a few years. Light Rail schemes are now actively being considered in other parts of Britain, including Manchester, Sheffield, Birmingham, Bristol and Edinburgh, in addition to further plans for London, notably in and around Croydon.

LOCAL RAILWAY HISTORY

About three quarters of the DLR re-uses old railway routes. Some of these older railways had been disused for many years, others remained in service right up to the day the DLR took over the tracks. The origin, development and, in many cases, decay of these railways is closely tied up with the history of the docks and each of the three arms of the DLR covers a different generation of railway development.

The City Route

In 1836 an Act of Parliament was passed authorising the construction of 'The Commercial Railway' running from Minories, by the City Wall, 3½ miles east to Brunswick Wharf at Blackwall. The railway passed close by the Regents Canal Dock, the West India Docks and the East India Docks. The railway was not initially intended to handle freight — rather it sought to attract large numbers of passengers whose journey from London to the Docks had until then to be made either by river — slow and often circuitous — or by road, which was even slower. Before the introduction of the telegraph, all messages had to be conveyed by hand and a continuous stream of clerks, message boys and businessmen travelled to and from the Docks daily. Additionally, the increasing popularity of the seaside had resulted in a growing number of steamer services from London to Kent and Essex resorts — services which could operate far more cheaply and efficiently if they started from Blackwall rather than from the Pool of London.

Two rival schemes had been put before Parliament for very similar routes to the docks and the unsuccessful rivals eventually merged with the Commercial Railway, with George Stephenson and George Bidder becoming the Company's engineers. In 1839, a year after construction had started, the company received parliamentary approval for an extension from Minories to Fenchurch Street and a formal change of name to The London & Blackwall Railway.

Opening on 4th July 1840, the London & Blackwall Railway was in its day a sophisticated and rapid system. Carried mainly on a 4,020-yard viaduct — the cheapest way of building in a congested urban area — the double-tracked railway was cable-hauled using a drum-to-drum system and seven miles of hemp rope for each track, with winding engines at either end of the line.

Within two years of opening the railway had extended into the City to Fenchurch Street and was experimenting with goods traffic. It was however isolated from the rest of the growing London rail network by virtue of its wide track gauge and cable haulage, the cables having a tendency to occasionally twist or snap, despite now being metal instead of hemp. To expand further the railway needed to standardise its equipment. In 1845 Parliament passed an Extension Act for the London & Blackwall Railway, authorising a connection with the Eastern Counties Railway at Bow and the change of gauge and haulage. The last cable-hauled train ran on 14th February 1849.

Poplar to Stratford

The Eastern Counties Railway was incorporated in 1836, to run from Shoreditch to Norwich and Yarmouth. Within the London area the Eastern Counties railway helped promote and then build the branch line from Stratford to North Woolwich in 1846/7 and had established at Stratford what was to become a major railway works. The London & Blackwall now became part of the growing Eastern Counties network, being eventually leased completely in 1865 to the Eastern Counties Railway, by now termed the Great Eastern Railway.

Other railways beside the Eastern Counties wanted to share in the lucrative docks traffic, and it was through a rival company, the North London Railway, that the DLR section from Bow Church to All Saints came to be built.

The North London Railway (NLR) started in 1846 as the East & West India Docks and Birmingham Junction Railway Company. The intention was to build a freight line linking the docks with the London to Birmingham line at Chalk Farm in North London. It took four years to build and open the line as far south as Bow, and it was not until 1851 that the railway reached Poplar. Poplar Dock was served by a large goods depot and an extensive yard of sidings with over 14 miles of track. Part of this area is now occupied by the DLR Operations and Maintenance Centre.

Passing through some of the more prosperous new suburbs of London, the railway company carried passengers from the start, although it was not until 1866 that passenger services extended south of Bow.

The heyday of the North London Railway occurred in the last twenty years of the 19th century. Over its Poplar line ran three other major railway companies' freight trains, and around Poplar Docks were grouped huge warehouses.

West India Quay to Island Gardens

In complete contrast to the NLR was the tiny neighbouring Millwall Extension Railway, the route of which is now used by the DLR from Crossharbour to Island Gardens. It began back in 1865 when construction work started on the Millwall Dock. Millwall Dock was built with an internal rail network, designed around horse-hauled wagons — steam locomotives were too much of a fire risk with quaysides of wooden ships, often with canvas sails.

At around the same time, the Great Eastern Railway and the Millwall Canal Company (owners of the Millwall Dock) jointly proposed a railway which would develop the southern part of the Isle of Dogs. Running south from Millwall Junction at the top of the Isle of Dogs, the line would skirt the east side of West India Docks and pass alongside the Millwall Dock to terminate on the bank of the Thames close by the jetty for the ferry to Greenwich.

Although this railway, the Millwall Extension Railway, would benefit its promoters, the neighbouring East & West India Docks saw the line as a threat, abstracting traffic, and they objected vigorously, especially to the proposed section through the West India Docks. Thus it was that the Millwall Extension Railway took six years to build, opening in 1871 to Millwall Docks and in 1872 to North Greenwich Station.

Single track throughout, with a passing loop at South Dock station, the railway was affectionately known locally as the 'penny puffer'. The 1880 locomotive and coaches survived until 1922 when three secondhand Great Western Railway railmotors (a steam engine and carriage combined) were bought as replacements.

The Twentieth Century

By 1926 few passengers used the Millwall Extension Railway — changing travel habits and increasing competition from motor buses bit deep — and in May 1926 the line closed to passengers. Competition from electric trams and motor buses was also affecting the London and Blackwall route from Fenchurch Street, and in the same month passenger services to Blackwall were withdrawn and the stations east of Stepney East closed. Only the North London Railway and Poplar line survived the inter-war years but the movement of freight continued and all three lines handled much dock traffic.

The railways suffered along with the East End as a whole from air attack in World War Two. The goods warehouses by Minories and the City were extensively damaged and the railway warehouses surrounding Poplar Docks were all but destroyed. Shadwell Station on the Fenchurch Street line closed to passengers in 1941 following bomb damage. The North London Railway Poplar line was crippled by a series of hits and closed to passenger traffic in May 1944.

Around the East End a great deal had changed after war ended in 1945. Many of the densely packed streets and houses had been destroyed and there was little reason to restore withdrawn passenger services. The lines remained open for freight and for the next 15 years continued to carry heavy traffic.

As traffic into and out of the docks declined, so the rail network was pruned back. First to go was the London & Blackwall line from Stepney East to Blackwall, closing in 1966. The Port of London Authority railways, which had included sections of the Millwall Extension, ceased in 1970. In 1980 both the West India and Millwall Docks closed. By then the North London railway route from Victoria Park to Poplar had been reduced to a single line and was becoming overgrown and derelict. Although not formally closed to traffic, services effectively ceased here after 1980.

Tower Gateway station site was squeezed in between a multi-storey car park (on the left) and Fenchurch Street BR tracks (on the right). New concrete decking was built, on which tracklaying is in progress in this November 1986 view.

BUILDING THE DLR

The DLR is a proper Act of Parliament railway established by the London Docklands Railway Acts of 1984 and 1985. The phrase Light Railway has been retained because of its association with Light Rail technology and staffing practices.

Of the two Acts, the 1984 one covers the City and Isle of Dogs sections and the Operations and Maintenance Centre, the 1985 Act the Poplar to Stratford leg. Although the first Bill was deposited in November 1982, it did not receive Royal Assent until April 1984, due partly to the 1983 General Election and partly to the hearing by Select Committee of the House of Lords of an unresolved Petition from a printing company, which successfully claimed that its sensitive equipment would be adversely affected by the railway.

The second Bill was deposited in November 1983 (Private Bills, such as those for the DLR, can be deposited only in November of each year) and received Royal Assent in April 1985 after Select Committees of the Commons and of the Lords considered Petitions on behalf of two separate groups of residents adjoining the railway at Bow. These were understandably concerned about visual intrusion, noise and vibration, even though very severe limits had been imposed in the contract awarded to GEC Mowlem.

At first sight it may seem as though the railway was blessed with being able to use a lot of redundant railway land that no-one else had previously claimed. However, almost the whole of the City arm has involved tenants trading from arches underneath the old viaduct; and at certain other places on the system, notably in the vicinity of Bow Church Station, other very sensitive property and occupation matters had to be tackled. Building a new railway on top of an old one, and sometimes beside an existing one as well, in a short time scale, with limited funds and often near people's homes, has been a task requiring skill, tact and resourcefulness.

Two-thirds of the 7½ route miles (12.1km) uses former disused or under-used railway lines. Even so, considerable new works — in volume and variety — have been necessary to accommodate the DLR. Beginning with the City route, the terminus at Tower Gateway is constructed on a new reinforced concrete viaduct. A double track viaduct has been constructed eastwards parallel to the British Rail Fenchurch Street lines. In this area a reinforced concrete slab supported by both the existing viaduct and independent foundations is used. Elsewhere an independent steel and concrete composite design has been used. From Cannon Street Road (1 km east of Tower Gateway) the DLR joins the BR viaduct and actually takes over the tracks of the former BR lines on the south side of the viaduct. The 113lb flat bottom rails remain, in contrast to the 80lb rail used elsewhere.

At Shadwell the existing viaduct carries a new island platform station built on the south side. A 200-metre reinforced concrete viaduct has been constructed south of the BR Stepney East station site to avoid the existing running lines and tie into the western end of the disused brick arch viaduct of the former London and Blackwall Railway. From here to the West India Docks the line uses more of the 1839-constructed viaduct. Generally, this structure was in remarkably good condition before work started: some arches required strengthening with a concrete overslab, others just repointing. Part of the viaduct is a grade II listed structure, in particular the 90ft span arches at Limehouse Basin. As well as repairs to the arches, 11 wrought iron bridge decks needed replacing with new concrete decks. To maintain the traditional viaduct elevation the original side girders have been removed, refurbished and put back in place again. At West India Dock Road the two-span bridge has been reconstructed to include the original solid pink granite columns in the road, for so long hidden under black paint. Between this bridge and the North Quay triangular junction the railway rises up from the brick arches using the standard steel and concrete composite structure to gain height. This junction incorporates three double junctions with 40-metre radius turnouts.

South of the junction the Docks Crossing begins. Although maximum use of standard rolled steel sections is made elsewhere, especially fabricated 65-metre spans are provided in each of the three docks with an eight metre clearance to the water.

South of the docks the line turns east on a typical Light Rail 50-metre radius curve and then winds its way south through the island on standard elevated structure, though a fabricated length is needed to cross the Millwall Cut water link between the Millwall and West India Docks.

South of Crossharbour the railway descends on to the earth embankment before being carried on a short length of standard new viaduct (containing Mudchute station) to tie in with the 27 surviving arches of the single track Millwall Park viaduct. Some repair and drainage work, raising of the parapet and new handrails have been necessary. The southern end of the viaduct has to be raised to provide clearance over Manchester Road. False arches have been incorporated into the new Island Gardens Station site.

The route from North Quay triangular junction to Poplar and Stratford contains several one-off features. The Docklands Northern Relief Road bridge is a 50-metre span skew plate girder bridge provided to cross a future four-lane road. The line descends rapidly towards Poplar station on a reinforced concrete viaduct over the top of the DLR electrical substation. Poplar station is constructed on retained fill to the west of the Operations and Maintenance Centre. The running lines turn northwards on a 60-metre radius curve, whilst the inside curve tracks to the workshops and stabling sidings include a 38-metre radius curve.

North of Poplar the trackbed has been re-ballasted and new drainage provided. A steel plate girder bridge has been built to carry the line over the Limehouse Cut Canal to replace the earlier badly decayed structure.

Bow Curve — taking the line from the old cutting to run beside BR on an embankment — features a 1 in 25 gradient and is laid on a 100-metre radius curve. New ballasted track was laid towards Stratford on the alignment of a disused line and minimal engineering work was needed to adapt the western end bay platform for DLR trains to use.

Spanning the West India Docks in 1985 caused a considerable rise in local land values as each one of the three docks was crossed. The view from South Quay looking northwards on 3rd June 1985.

The north route being graded prior to casting of the track bed on 19th May 1985. This picture is taken between Bow Curve and Bow Church station looking south.

The lightweight construction of the viaduct is clearly shown in this view from Millwall Cut looking west towards the site of South Quay station before the casting of the concrete deck in 1985.

Island Gardens station is on the site of North Greenwich station which closed in 1926. This June 1985 view taken from the end of the Millwall Viaduct shows how the original station was totally demolished and a new concrete and brick structure begun.

February 1986 and the contractor's tracklaying train has just brought a section of track up from the temporary fabrication site — now occupied by the depot trainwash — to the cutting south of All Saints station. The track-laying machine has lifted the track section and running on temporary rails will carry the section into place.

After the Royal opening of the railway on 30 July 1987 Her Majesty the Queen, here accompanied by LRT Chairman Sir Keith Bright, rejoins car 11 at Poplar for the run to Tower Gateway.

DLR TRAINS

On any railway the trains present the most important element of a passenger's journey and the most visible to the casual observer. It is generally recognised in the 1980s that a train is not solely a functional piece of hardware, but an essential part of the image both in external appearance and internal finish and comfort. This was recognised by the railway's joint clients and by the contractor, and substantial effort was made to ensure high standards of engineering and finish.

With modern technology, the train controls become an integral part of the operating system. Consider the need for the train to react to the three elements of the signalling system to ensure a safe, timely and accurate passage from one station to the next and also the requirement for the train to communicate status changes, such as passenger alarm operations or change of mode of driving (eg into manual). Looking more closely at safety responses, the car doors, control panels, power circuits and braking systems all have direct communication links with the 'outside world' of signalling commands in addition to the internal connections normally found. With this in mind it is clearly not practicable to simply buy a train off the shelf, put it on the rails and expect it to function. More than ever before, the rolling stock has to be seen as an integral part of the overall system.

Light rail vehicles (LRVs) are produced in quantity in Germany, France and Italy. West German technology in particular is exported across the world, notably to North America where, like Britain, the lack of a domestic market for years inhibited the development of a home-based product. DLR's contractors were thus faced with either opting for overseas designs or starting from scratch in Britain. 'Re-inventing the wheel' is at best a wasteful concept and, in the context of DLR, evolving a whole new design for just 11 cars could not be justified. Main contractor GEC-Mowlem therefore co-operated with Linke Hofmann Busch (LHB), one of West Germany's main builders of main line passenger and freight stock, metro cars and light rail vehicles.

The absence of a driving cab lets passengers take the 'driving seat'; this has proved popular with people of all ages. As on the more recent Underground stocks, passenger 'Door Open' facilities are provided to prevent unnecessary heat loss in wintery conditions (the operating button is visible to the left of the doors).

The cars were delivered to Poplar Depot by road and 01 arrived on 7th August 1986. This view was taken on the following morning, when with the use of two mobile cranes both halves were lifted onto the track and joined together.

During commissioning trials vehicles ran an average of 18,500km, trials culminating in a full 'ghost' service during August 1987. Here Car 11 enters Bow Church station.

Inside the vehicle the open layout allows the Train Captain to move about easily. Traditional London moquette seat covers are used in preference to the European standard plastic 'bucket' seat.

The front windscreen is one of the toughest parts of the vehicle. One early problem was the electric destination blind mechanism which was not successful and resulted in caught and torn blinds being replaced by temporary destination boxes on some vehicles.

MAIN FEATURES OF THE VEHICLES

GENERAL
Two-body articulated units of the following dimensions:

Length: 28m (91ft 10in)
Width: 2.65m (8ft 8in)
Height: 3.40m (11ft 2in)
Weight: 39 tonnes (38.4 tons) approx (tare)
Bogie centres: 10m (32ft 10in)
Bogie wheelbase: 2.10m (6ft 11in)
Min track curve radius: 40m (130ft)

Passengers: 84 seats; total capacity 210 with normal standing load and 260 with crush load.

Capable of operating singly or as multiple-unit trains with automatic coupling.

Maximum design speed 80 km/h (50mph).

'P86 stock' (Cars 01 to 11) built 1986 by Linke Hofmann Busch, Salzgitter.
'P89 stock' (Cars 12 to 21) to be built 1988-9 by BREL, York.

BODY
Steel construction with polyurethane painted finish.

Four doorways each side per car, each of 1300mm clear opening, fitted double inward-gliding doors, electro-pneumatically operated passenger control of door opening.

Seats: glass reinforced plastic frames with moquette covered cushions and squabs.

BOGIES
Of welded steel H-frame construction, with rolling rubber ring primary and airbag secondary suspension. Roller bearing slewing rings.

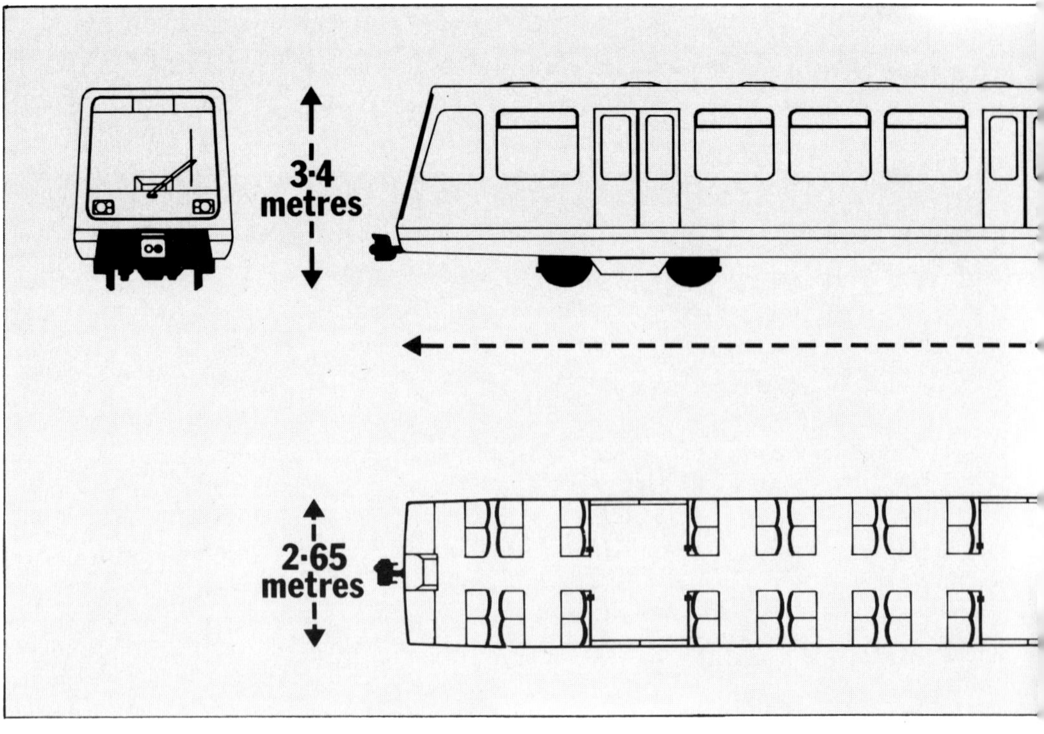

The outer bogies are motored, the longitudinal frame-mounted motor driving both axles via right angle gearboxes and flexible drives.

Single disc brake per axle, the actuators on the motored bogies incorporating spring applied parking brakes.

Resilient wheels of 740mm (2ft 5in) diameter new, 660mm (2ft 2in) minimum.

TRACTION CONTROL
Supply: 750 volts dc from an under-running third rail.

The two motors are fed in parallel from a two-phase chopper giving an overall frequency of 538 Hz. Each half of the chopper uses a single Gate Turn-off Thyristor (GTO).

The motor fields are series fed in motoring, but separately excited in rheostatic braking. Regenerative braking is not provided.

Maximum accelerating current: 700 amps/vehicle.

AUXILIARIES
24 volt battery charged by ac-dc converter

Compressor: Hydrovane (rotary vane) type TB11/C14.

Automatic couplers: Scharfenberg

PERFORMANCE DATA
Acceleration: 1.0m/S^2
Service braking: 0.8mS^2
Emergency braking: 1.2mS^2

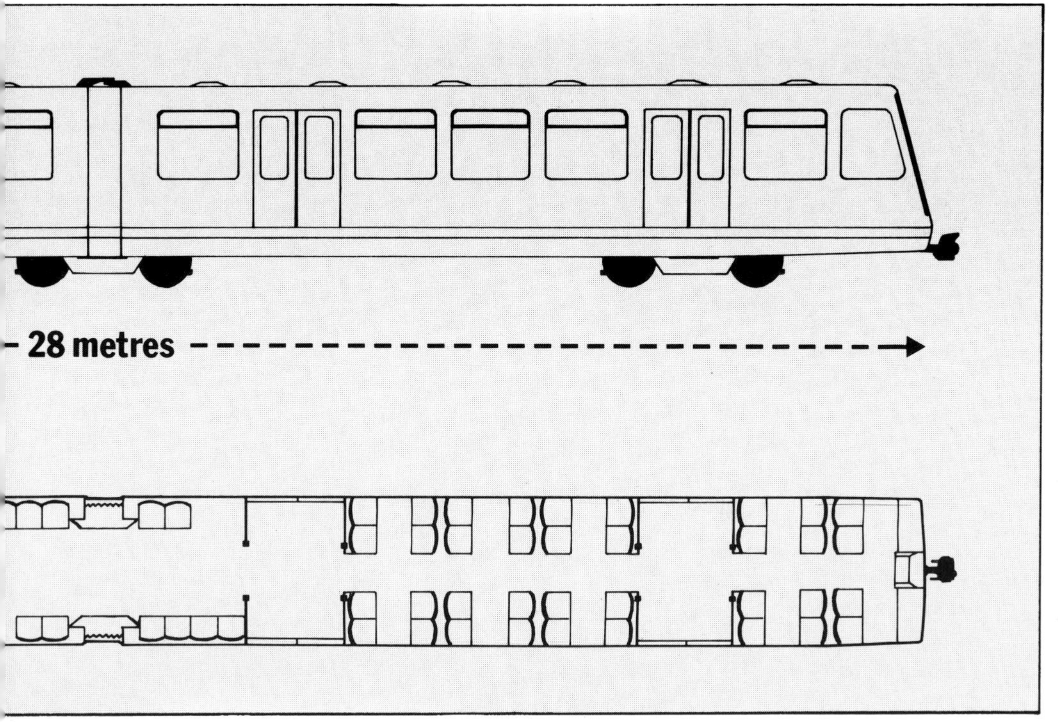

Maintenance facilities are provided by a Wickhams CT30 vehicle — known to all staff as 'Sooty' and now named as such. Sooty was used extensively during the later construction stages and is seen at Island Gardens in Febraury 1987.

Car 11 was delivered to Manchester first, and fitted with overhead current collection equipment it ran for two weeks in March 1987 on a disused section of British Rail track at Debdale Park. Rewired with lightweight overhead power lines and with a temporary station, this section was used as a demonstration to Manchester of the potential of light rail.

Under the agreement reached, LHB were to be responsible for the shell and bogies, whilst GEC would fit all electrical components in Britain. In the event, a subcontract to have the interiors fitted out by one of Britain's leading builders failed to materialise and the cars were completed at LHB's Salzgitter works, with the first one rolled out with some ceremony on 20th June, 1986. The 11 were delivered by road and sea via Hamburg and Kings Lynn between August 1986 and February 1987. The last car — No. 11 — was to make history twice, firstly by being delivered direct to Debdale Park, Manchester as the key part of a light rail demonstration where it became the first revenue-earning DLR car.

The car was modified to accept a pantograph for the overhead power collection system installed by Balfour Beatty, a conversion which was remarkably simple involving little more than an additional roof member and a length of conduit to carry the main power cables down to the underframe.

Car 11's main clame to fame, however, was to be on 30th July when, in the hands of Train Captain Gary Bonini, it carried Her Majesty the Queen and His Royal Highness the Duke of Edinburgh from Island Gardens to Poplar, and thence into Tower Gateway, following the Royal Opening of the railway.

As mentioned above, great care was taken with the interior and exterior design of the cars. The seats were mostly arranged in bays of four — always popular on the Underground and elsewhere — to take full advantage of the large windows offering fine rooftop views of the changing docklands scene. Nowhere is this more true than in the end seats, giving passengers an almost unprecedented driver's eye view. Understandably these seats are almost always the first to fill. To increase circulation space longitudinal seats are provided in the centre section of the train, along with two wheelchair bays. The railway has some regular wheelchair users, but the bays are most heavily used by shoppers with wheeled trolleys and baby buggies taking advantage of the railway's superb roll-on/roll-off facilities.

In order to achieve good access for unassisted wheelchair users the cars have been designed to have a level car floor to platform gap of only 75mm (approx 3 inches). This ruled out conventional exterior plug doors and led to the adoption of inward-opening swing plug doors. These were never an ideal solution but with the traffic levels forecast at the time a contract was awarded it was anticipated that the layout would be acceptable. The enormous upturn in development on the Isle of Dogs has put this aspect in a different perspective, however, and future orders are likely to have an alternative arrangement.

With the cars operating as single units, reliability has to be a major concern. DLR does not have the luxury of multiple power cars to enable a train to 'limp home' on half power in the event of a single failure. Breakdowns certainly have occurred — notably as a result of problems with the current collection shoegear and, in the early days, of some remaining software bugs — but complete stoppages have been gratifyingly few and in most areas co-operation between the operators and suppliers produced rapid improvements. The prophets of doom looked at earlier generations of rolling stock and forecast that with single cars DLR would never complete a day's running. Well, admittedly DLR's policy of training its staff to deal with 'pushouts' has been vindicated, but the woeful predictions were very wrong!

The initial fleet of 11 cars is being supplemented by a further ten to achieve the capacity needed for the first phase of services to Bank. These cars will be very similar to the initial batch with only detailed modifications — mostly under the skin — made in the light of operational experience and the need for longer trains. To meet the needs of the full upgrading a further ten trains will be ordered during 1988. These may have more fundamental changes to reflect the higher loads expected, and it is intended that they should be the forerunners for a fleet for the Beckton extension — time will tell!

The ten cars ordered as part of the upgrading contract in July 1987 will, however, differ from the initial cars in that they will be constructed in the UK. During 1986 BREL decided to enter the light rail market and entered a co-operative agreement with LHB, whose products are largely complementary to those of the UK builder. The DLR batch will be the first products of this development, but this should be only the beginning.

THE ROUTES DESCRIBED

Geographically the DLR consists of three legs joined at a triangular junction, which is located to the north of the former West India Docks system. The three legs terminate in the City (at Tower Gateway, but later Bank station), Stratford to the north and Island Gardens, a terminus at the southern end of the Isle of Dogs peninsula facing Greenwich.

Car 07 passes between the Regents Canal Dock and St Anne's Church, Limehouse. DLR now has exclusive use of this brick arch viaduct built in 1840 for the London and Blackwall Railway but abandoned by British Rail in 1962.

THE CITY ROUTE

Tower Gateway terminus is built on a new structure close to the eastern end of the British Rail Fenchurch Street platforms. The DLR terminus is on the east side of the street called Minories. Nearby Tower Hill Underground station provides interchange between DLR services and the Circle & District lines of the Underground.

From Tower Gateway a new double-track structure takes the line east as far as Cannon Street Road. Between this point and Stepney the DLR moves on to the existing viaduct where British Rail has given up two of its former tracks. A station is located at Shadwell, where the viaduct crosses Watney Street, for passengers to change for the East London Line Underground station in nearby Cable Street, or to reach Watney Market shopping centre to the north of the station.

Near Limehouse, by the British Rail Limehouse (formerly Stepney East) station, a new structure is provided to carry the line over Butcher Row. A new DLR station is built to the east of Butcher Row, allowing interchange between British Rail and DLR services. East of this station the line leaves the BR alignment and follows the course of a disused railway viaduct. Although this structure was built as long ago as 1840 for the London and Blackwall Railway, it is still in good condition.

The line passes close to the Regents Canal dock — also called Limehouse Basin — and then continues east crossing Westferry Road where another station, called Westferry, has been built. Then, after crossing West India Dock Road, the site of the North Quay high level triangular junction is soon reached, just to the north of the West India Docks, the final ascent to the junction being a spectacular 4.3 per cent gradient.

Limehouse station is an increasingly important interchange between DLR and British Rail (London, Tilbury and Southend Line), whose platforms are visible on the converging route beyond car 04. With the building of DLR, extensive alterations were carried out to the BR station, giving the two a combined entrance. The BR station was known as Stepney East until May 1987.

A well-loaded Car 10 draws into Shadwell station. On the right are the two British Rail Fenchurch Street tracks. The heavyweight overhead electrification on the BR line contrasts with the slim unobtrusive third rail system of the DLR.

Westferry station was built over two road junctions with relatively little pavement space being available for the DLR to use. The canopy, lift and stair arrangements have had to be staggered but all the DLR station elements are included.

THE ISLE OF DOGS ROUTE

The route through the Isle of Dogs to Island Gardens runs from the North Quay Junction, on the north side of the former docks, to West India Quay station, which is built in the space between two quayside warehouses on the former West India Import Dock. A series of new bridges then carries the line across all three docks in the West India system, with a station at Heron Quays — Canary Wharf station will open later.

After crossing the docks the route curves sharply to the east and descends slightly to South Quay Station before rising again in order to cross the Millwall Cut, again with a clearance of 8 metres (26 feet) above water level. From here the line structure, passing Crossharbour station near the Asda store at the District Centre.

After this station the alignment of a disused railway embankment carries the line south beside East Ferry Road to a point where it turns east and crosses the road on a new bridge which also supports Mudchute station. Leaving Mudchute station, the construction continues on to another old viaduct. This structure is a single-track viaduct built in 1872 and is still in good condition despite the fact that it has been totally disused since 1926. This 27-arch viaduct runs through Millwall Park, ending at Manchester Road. Here a new bridge carries the line across Manchester Road to a two-platform terminus at Island Gardens. Passengers can cross the Thames to Greenwich using the nearby 1902 pedestrian tunnel.

The line to Stratford heads east from the North Quay triangular junction and rapidly descends to Poplar station. The Operations and Maintenance Centre is on the northern side of the line, northeast of Poplar station. The route then turns into shallow cutting to follow the course of another disused railway to Bow.

Trains are carried across the former West India Docks on striking 65m long spans. The fine long entrance views currently afforded will soon be contained by spectacular development on Canary Wharf (to the right) and Heron Quays.

A Stratford to Island Gardens train negotiates the lofty North Quay Junction. The route westwards to the City can be seen descending the 1 in 17 gradient into the distance beyond. Tight curves — up to 40 metres radius — have enabled land take to be reduced to a minimum; future development will be built up to the railway here.

South Quay Plaza, beyond the railway at South Quay station, became the first major Isle of Dogs development to be occupied when Daily Telegraph editorial staff moved in during the first half of 1987.

At Heron Quays it is intended that the station, paid for by the developer of Heron Quays, will eventually be encased in a larger scale version of the unusual triangular buildings on the right, giving undercover access to most of the development.

From Mudchute to Island Gardens the DLR can take on a surprisingly countrified look in the summer. The topmasts of the Cutty Sark and the twin towers of the Royal Naval College emphasise the closeness of Greenwich, just across the River Thames.

Stations have been built at All Saints, Devons Road and Bow Church and there is provision for an extra station at Carmen Street should finance become available. North of the Bow Church stop the railway is reduced to a single track and a new climbing curve carries the track from the cutting of the former North London Railway to the embankment of the high level British Rail Fenchurch Street to Stratford line.

From here the line makes use of a surplus single BR track at the south side of the Liverpool Street to Stratford Main Line. An additional stop at Pudding Mill Lane between Bow Church and Stratford is also under consideration. Fortunately a bay platform, never used by normal services, is available at the western end of Stratford's up platforms, and this is used by DLR trains. Passengers can change to the Central Line and BR services, as well as leave the station to reach Stratford shopping centre, market and bus station.

Before and After. Between All Saints and Devons Road stations is the site of the possible Carmen Street station. These two views, one taken in August 1984 and the other in January 1987, show the progress made. The footbridge over the railway is all that remains of South Bromley station which, with its island platform, can just be discerned in the 1984 view.

At Bow Curve the DLR narrows to a single track and climbing at a gradient of 3% and a curvature of 95m radius, rises from the old North London Line to meet and run with the main East Anglian line from Liverpool Street.

Stratford station gives cross-platform interchange to the westbound Central Line and the BR local line to Liverpool Street as well as undercover access to the bus station and shopping centre.

POWER SUPPLIES

The DLR obtains its main power supplies from two separate London Electricity Board (LEB) intakes at Poplar and through a substation located under the railway west of Poplar station. This substation feeds separately each track of each of the railway's three branches. Power for signalling and the Operations and Maintenance Centre is also obtained from these sources. The duplication of intakes and of substation equipment is intended to guard against total loss of supplies. However, in the event of this happening, essential lighting and communications circuits are maintained to permit safe and orderly evacuation of the railway. Station power supplies are obtained locally from the LEB street network and, again, a battery back-up system is provided to maintain an adequate level of lighting for safety.

Traction current is distributed at 750v dc by means of a small-section aluminium conductor rail with a stainless steel contact surface bonded to the body of the rail. The superior conductivity of aluminium compared to conventional steel rail, and the low current demands of the vehicles, permit the use of a single substation to feed the entire system. The current rail is of bottom contact type with a plastic shroud protecting the upper and both side surfaces, minimising the risk of accidental electrocution and avoiding the problem of snow and ice normally associated with low-level current collection systems. The trains themselves are fitted with air-applied copper collection shoes on glass reinforced plastic arms, mounted centrally on each side of the motor bogies.

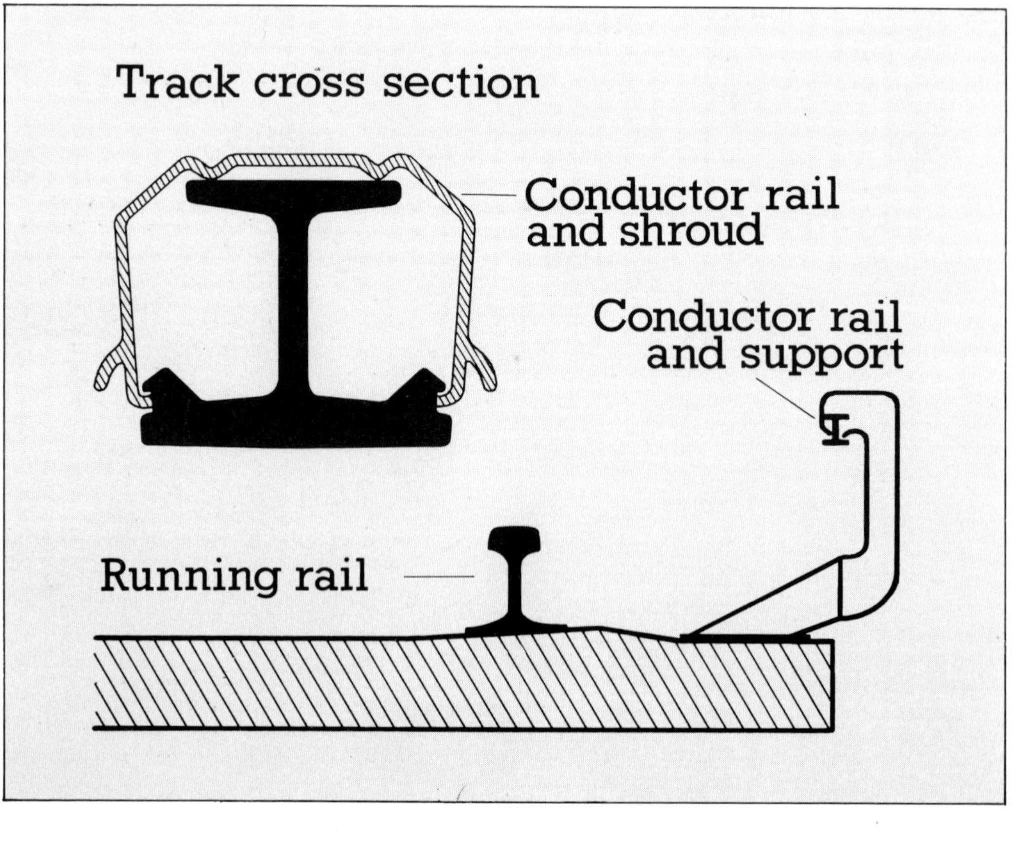

PERMANENT WAY

It was originally planned to use a low-maintenance non-ballasted trackform throughout the system, but an extensive noise study demonstrated that ballasted track would substantially reduce noise levels and greatly benefit communities adjoining the railway. A concrete trackbed has been incorporated on all new elevated structures to keep to a minimum the imposed load and consequently to contribute to the economy of the design. In addition, it was specified that all curves less than 100 metres radius should be fixed to a concrete trackbed.

The standard rail cross section adopted for the system is BS80A per yard flat-bottom rail. The Pandrol 'e' clip fastening is used with adapted British Rail F24 type sleepers for ballasted track and a cast iron baseplace assembly on slab track. The fastening assembly incorporates 10mm thick rubber-bonded corkpad between rail and baseplate or sleeper. Continuous welded rail is used throughout with alumino-thermic welds. At a location sensitive to vibration, where flats at Holyhead Close, Bow, had previously been built on a raft deck over the line, James Walker sleeper soffit pads are installed.

As a result of the tight railway geometry, 40-metre radius turnouts have been specially designed for the system. Where possible standard BR Cv 9¼ turnouts (adapted for 80A rail) are used.

RAILWAY HEADQUARTERS

Not surprisingly, this local railway has local headquarters in Poplar between Poplar High Street and the New Billingsgate Fish Market. The headquarters, the full title of which is the Operations and Maintenance Centre (OMC), contains all the administrative, operational and maintenance functions necessary for the running of the DLR.

The OMC is not designed to provide full heavy overhaul facilities for the small DLR fleet, concentrating instead on service maintenance with some overhaul and repair capacity. It is intended that the workshops could carry out all work up to a half-life overhaul, with selected items and certainly some larger items sent away for repair and refurbishment.

The steel-framed building built on piled foundations is finished in a distinctive blue profiled steel cladding with the reception and first floor control room located at the western end; all administrative and operational offices are located on first floor level above ground floor workshops and plant rooms. At the eastern end are the vehicle workshops, complete with three parallel tracks with sunken pits. The northernmost of the three pits is built around a so-called swimming bath pit that is substantially wider than the track gauge to allow for exterior low-level bogie inspection. There is an electrically-operated 7.5 tonnes capacity overhead crane at the western end of the workshops operating at right angles to the lie of the pit tracks and extending out towards the south side of the building where there is road vehicle access.

Five stabling sidings are provided on the north side of the OMC. These feed eastwards into two transfer tracks, which lead to the running lines either towards Poplar or towards All Saints. The latter access traverses the train wash which is placed close to Poplar High Street overbridge.

To the north of the stabling sidings are the fenced permanent way stores and a large earth mound which is now being removed to make way for more sidings between the train wash and the north east corner of the workshops area.

The office accommodation within the OMC proved inadequate for the operating team and for the designers and planners involved in contract monitoring and extension planning, so temporary site offices in Stoneyard Lane (300 yards west of the OMC) have been retained for these staff, as well as for training new DLR staff. GEC-Mowlem's headquarters offices are also located here.

Poplar OMC on 4th May 1987. To the left is the delta junction, whilst on the right the Stratford branch runs into cutting to All Saints station and on up towards Devons Road station. The Beckton Extension will come off the curve around the depot.

The OMC sidings are the only area on the DLR where trains are normally driven in manual to conventional colour light signals. On the extreme left is the bogie wash where vehicle bogies can be hosed clean using high pressure equipment prior to maintenance work.

On the first floor at the front of the OMC is the Control Room. The left wing of the desk deals with power and with train radio, the centre with automatic train control/signalling and station platform monitors, and the right wing with p.a., platform describers, emergency alarms and telephones.

SIGNALLING AND TRAIN CONTROL

It is widely known that DLR trains are 'automatic'. But just what does this mean? There are in fact three overlapping automatic systems — Automatic Train Operation (ATO), Automatic Train Protection (ATP) and Automatic Train Supervision (ATS), with the respective functions of driving trains, ensuring safety and timekeeping.

Automatic Train Operation is essentially the driver. Like a human driver it is located on board the train and, based on information given to it by a transmitter at each platform called a Docking Data Link, it can recall from memory exactly where to accelerate, coast and brake to get to the next station. Speed restrictions for sharp curves are taken into account and two separate speed profiles can be chosen — one with coasting for economical operation, the other an all-out dash to regain lost time. The information which determines the profile comes from the ATS package. ATO routes are available on all running lines and to and from the OMC, but not within the depot or over emergency crossovers.

Automatic Train Protection ensures that a train cannot go too fast for the section of track it is traversing or enter a section already occupied by another train. Provided the ATO system does not make a mistake (nor, in the event of manual driving, does the driver) this system lies dormant, like a policeman, waiting to step in before any error leads to an unsafe situation.

On the track, the system operates by the train detecting a code transmitted from two cables — one the 'safety tone' which is absent when a train is already in the section, the other the 'speed tone', which has loops laid at varying distances such that they correspond to one second intervals at maximum line speed for the location. If a train detects more than one loop per second, or misses one, it will stop. By operating a push button in the control room, the supervisor can immediately stop the trains in a selected area in the event of emergency — for example a person falling onto the track.

The co-ordination of the safety features, including points locking, is carried out by Solid State Interlocking (SSI), a failsafe electronic equivalent of conventional mechanical and electrical locking used since railways began. The SSI used by DLR was developed by British Rail and GEC and specified following a successful pilot installation at Leamington Spa.

Again, the Docking Data Link has an important part to play in ensuring correct door operation at platforms and correct direction operation by the train.

Automatic Train Supervision provides the means by which the train service is regulated. This can automatically adjust departures to regain lost time, but is capable of providing manual control of the system through computer keyboards. This facility allows the supervisor to either make commands specific to particular equipment, or make more generalised inputs — for example to add an additional train journey, which will then operate as required without further intervention. It is also possible to write an entire new timetable in-house, which can be operated as required at the start of a new day — or even part way through.

The ATS system also drives the platform train indicators and the station public address, but as with the other automatic systems these can be overridden manually if required.

Whilst the Docklands installation is in many ways breaking fresh ground, many of the component parts are tried and tested elsewhere — SSI at Leamington Spa, ATP on the Birmingham Airport Maglev and ATS in Buffalo, New York and in Atlanta, Georgia.

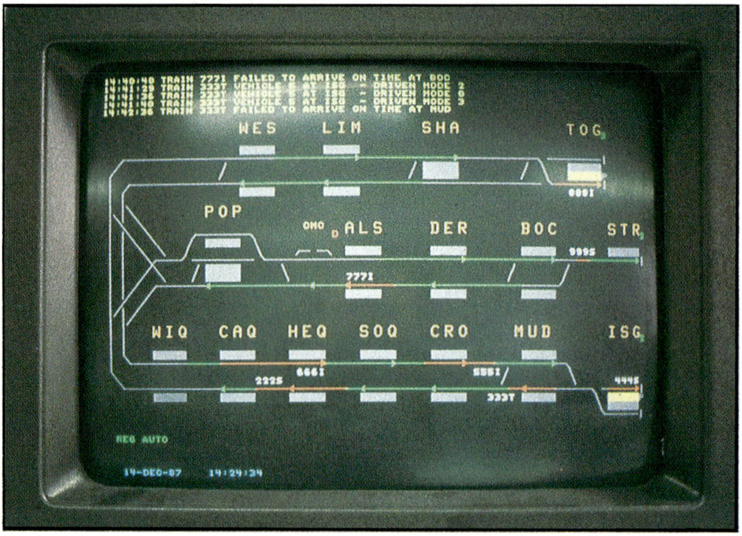

An overview layout on the control screens displays the whole railway with vehicles as red bars and routes in green. Yellow platforms indicate a vehicle has 'docked' at a station and the control computer can communicate with the vehicle computer, whilst at the top of the screen the computer notes variances from the automatic timetable.

The three track workshop at the rear of the OMC is for running maintenance and light to medium overhauls of vehicles. Six co-ordinated electric trolley jacks allow a vehicle body to be lifted from the bogies, the latter then being moved via small turntables and adjacent tracks to the heavy store area at the front of the workshop beneath an overhead hoist.

PASSENGER TIMETABLE

The philosophy behind the passenger timetable is based on the notion that the service should be so frequent and consistent that passengers hardly need to consult conventional timetables. The standard timetable operates between 7am and 7pm, Monday to Friday. A reduced service operates before 7am and after 7pm to midnight, and at weekends.

Although the Performance Specification to which the railway was designed, specified eight trains an hour on two services (City and Stratford), the railway opened with six trains per hour on each service. Testing to ensure satisfactory operation of the full Performance Specification took place during 1987, with a view to stepping up the service in 1988 to that originally planned for public opening.

In practice, the standard timetable is changed according to changing circumstances. Occasionally the weekend service requiring seven cars has been operated like a Monday to Friday service (when nine cars would normally operate). On some Sundays an extra vehicle has operated Tower Gateway – Island Gardens shuttles to satisfy remarkably heavy passenger loadings.

FARES AND TICKETS

The DLR fare structure is in line with the London zonal fare system as applied to the Underground, but the ticket system has many features more in common with Swiss or German tram operation than with traditional heavy urban railways in Britain. The DLR policy can be described as hybrid: tickets are not sold on the cars themselves but casual passengers buy their tickets in machines before boarding a train, as on the Underground, and tickets can also be bought from local shops. All single journey tickets which are of the Docklands pattern (they do not look like traditional Underground tickets) need to be validated (or cancelled or encoded) before travel. The validation process is carried out in a blue totem pole strategically positioned between ticket machines and entrance to the station. Validation consists of code marking the ticket with such data as location (station name, date and time) to make it valid. The validators not only imprint a visible statement but also a magnetic encoding on the back of the ticket.

All DLR stations are open — there is no barrier — but a red line always appears on the way up to or down to the station platforms. Like some German tram subways, passengers can be charged fixed penalty fares if caught beyond the red line without a valid ticket. The trains are staffed with the popular Train Captains; amongst their duties ticket inspection and revenue protection are the most important. Additionally, roving inspectors and staff in plain clothes are used for spot ticket checks, with British Transport Police and Metropolitan Police backup.

Passengers with valid tickets issued at Underground or Network SouthEast stations, Travelcards, Capitalcards, senior citizen travel permits, passes and other seasons valid for the zones through which the DLR passes, can go straight to the platforms as their tickets are regarded as already validated. The reverse is also true, so that passengers with DLR tickets bought at DLR ticket machines can continue their journeys by Underground. Docklands tickets will be fully compatible for single journeys trips with the future Underground Ticket System (UTS) — then DLR tickets will open automatic gates at Underground stations.

Most DLR stations are in Inner Zone 2, the only exception being Tower Gateway in Central Zone 1 and Stratford in Outer Zone 3a. The ticket machines and validators have been supplied by Thorn-EMI Electronics, Wells, Somerset. Tollpoint ticket machines and Travelog ticket validators have been supplied. The ticket machines are linked to a computer in the control room at Poplar and the status of all machines is recorded automatically, showing control room staff when the machine is out of paper roll stock, out of change or full up with money. The Travelog validators take in the whole of the DLR ticket to read it, rewrite the code print on the front, then return it to the customer. None of the ticket is cut off at the time of printing as on many Swiss/German validators where, incidentally, passengers hold on to one end of the ticket. DLR tickets disappear into the validator and are returned in a matter of a couple of seconds. The Travelog validators are housed in distinct totem poles and the vending machines are mounted in substantial brick buildings which are alarmed and sensitive to such things as changes in air temperature caused by unauthorised intrusion.

It is true to say that the greatest single cause of disappointment to railway management — and the greatest annoyance to the passengers — to date has been the performance of the ticket issuing and validation equipment. It was always recognised that it would be a difficult proposition to not only launch a new railway with unfamiliar locations but also with unfamiliar ticketing and operating habits. However, the actual performance of the machinery has led initially to multiple complaints, confusion and extreme passenger annoyance in some cases, through its basic unreliability. The original problem of ensuring that people validated Docklands tickets turned out not to be such a problem! In fact, passengers attempted to validate Travelcards, Capitalcards, staff passes and other tickets, which caused validators to go out of service in the first few weeks. However, the process of accepting coinage and giving change with the ticket largely defeated the Thorn-EMI Tollpoint machines initially and, more worryingly, only certain faults were clearly reported back to the control room computer.

Had the ticket issuing machinery worked effectively there would have still been enough problems with the speed of issue of single journey tickets and multi-value, multi-choice decisions that passengers have to make individually at the one machine. These reasons on their own would have caused some initial delays but were compounded by the sheer number of failures to provide a ticket and/or change — or even more annoying to passengers, to refuse to accept perfectly legal coinage in the first place! A number of supplementary ticket-issuing methods were adopted from the first day of public opening and these included batch-issued standard tickets supplied from a remote ticket machine

Left **Pairs of ticket machines are mounted in blue brick buildings at each station. Remotely monitored from the Control Room at Poplar OMC; the tickets issued, coin held and current status of each machine can be checked using the ticket computer terminal.**

Right **Tall blue validation columns adjacent to each platform entrance must be used by holders of pre-purchased or machine-issued single tickets to validate the ticket with the station name, date and time, prior to travelling.**

in the head office at Poplar. Tickets issued by this machine can be sold by authorised managers or any spare traffic staff from any given location. Such tickets have the machine code 000 printed rather than the three-figure numerical sequence codes of the station-installed Tollpoint machines. In practice, such tickets have been sold from Tower Gateway and the Island Gardens concourses and by post to enquiring school parties. Additionally, hand-written London Underground excess fare book tickets were issued from the afternoon of 31st August and have been issued intermittently since to solve local traffic pressures. These books have now been replaced with DLR Additional Fare books as used by Train Captains on the trains. Such tickets are issued by hand to reduce waiting times, particularly at weekends at the termini, or gateway, stations.

The first serious local promotion ticket — the 'Docklander' — was experimentally issued from the Island Gardens exhibition shop on Saturday 27th September. This is a ticket priced between the cost of two single-journey tickets between Island Gardens and Tower Gateway and return and a standard London one-day Travelcard. At £1.80 adult and 65p child it provides all day travel at will and break of journey rover facilities on the DLR. As such, it is aimed almost entirely at the surprisingly large number of passengers arriving at the Island Gardens end of the railway from Greenwich who wish to have a sample ride on the railway. It was thought that most tourist traffic would originate at the Tower end of the railway, where the annual attendance figures at the Tower of London and associated attractions outnumber two to one the present figures for attendance at Greenwich. However this does not seem to be the case and a significant market segment has been targeted with the Docklander ticket sold at Island Gardens. Such tickets can be sold from other concourses at other times according to passenger demand and staff availability. An additional ticket not often seen by the travelling public is a Visitor's Day Pass which is issued by the DLR to company guests and has a similar validity to the Docklander ticket. The Docklander ticket does not need to be validated and is made to a deliberately larger size to prevent this! The Visitor's Day Pass, on the other hand, is the same size as a standard DLR ticket with a dummy magnetic code on the reverse. It only requires counter signatures to validate and does not require separate electro-mechanical validation.

Experience with fare checking and dealing with fare evasion shows that this part of the DLR policy has been an overwhelming success; attempts at sustained fare evasion have been largely squashed. Initially a lot of cases occurred where there was some doubt about the performance of the ticket issuing equipment, leading not surprisingly to a number of altercations on the trains. However the roving presence of the Train Captain and the obvious visible proximity of back-up staff, also radio equipped, and British Transport Police mean that to date this part of the ticketing policy has been successful.

Docklands

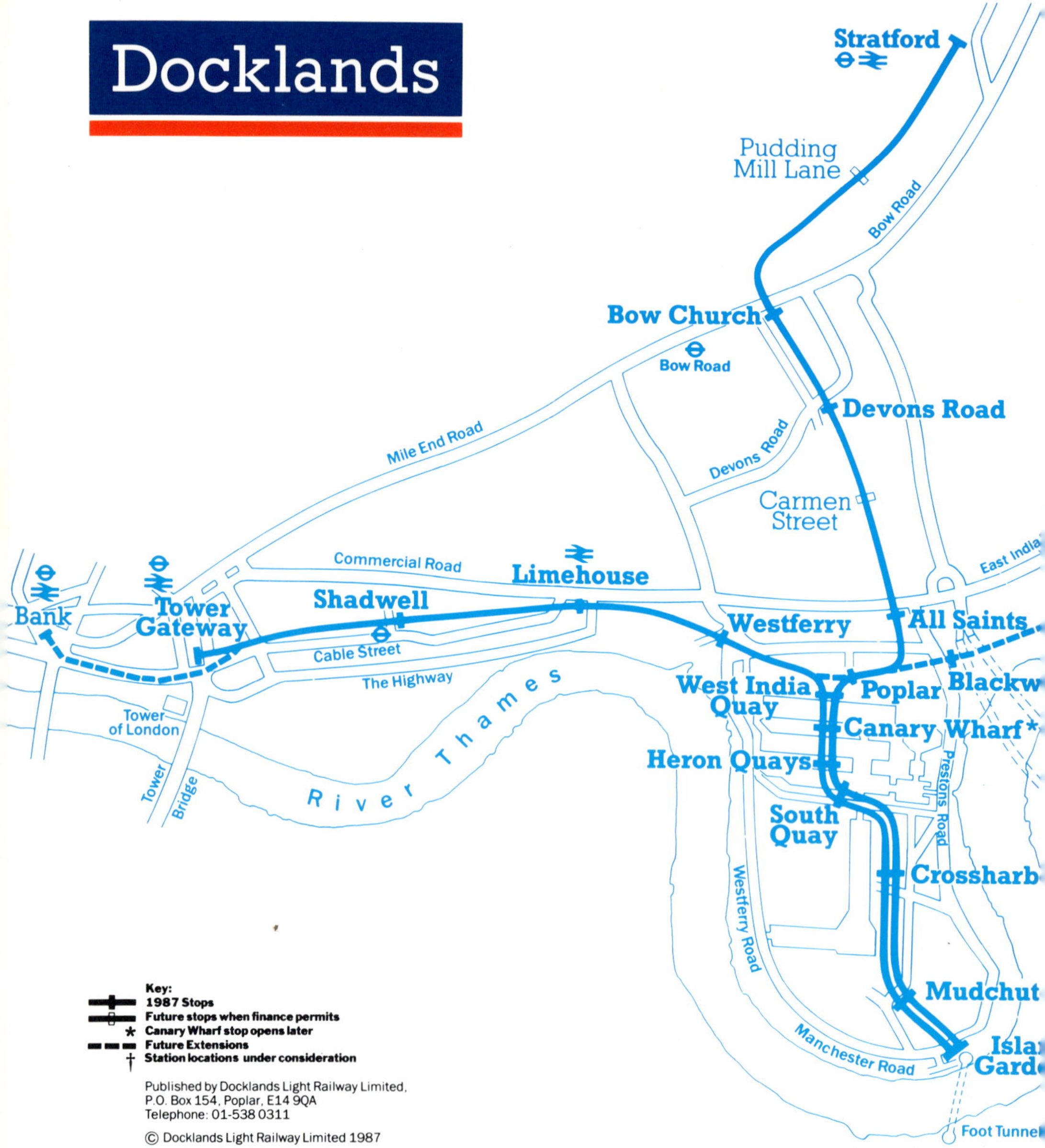

Key:
1987 Stops
Future stops when finance permits
* **Canary Wharf stop opens later**
Future Extensions
† **Station locations under consideration**

Published by Docklands Light Railway Limited,
P.O. Box 154, Poplar, E14 9QA
Telephone: 01-538 0311

Map No. 13 March 1988

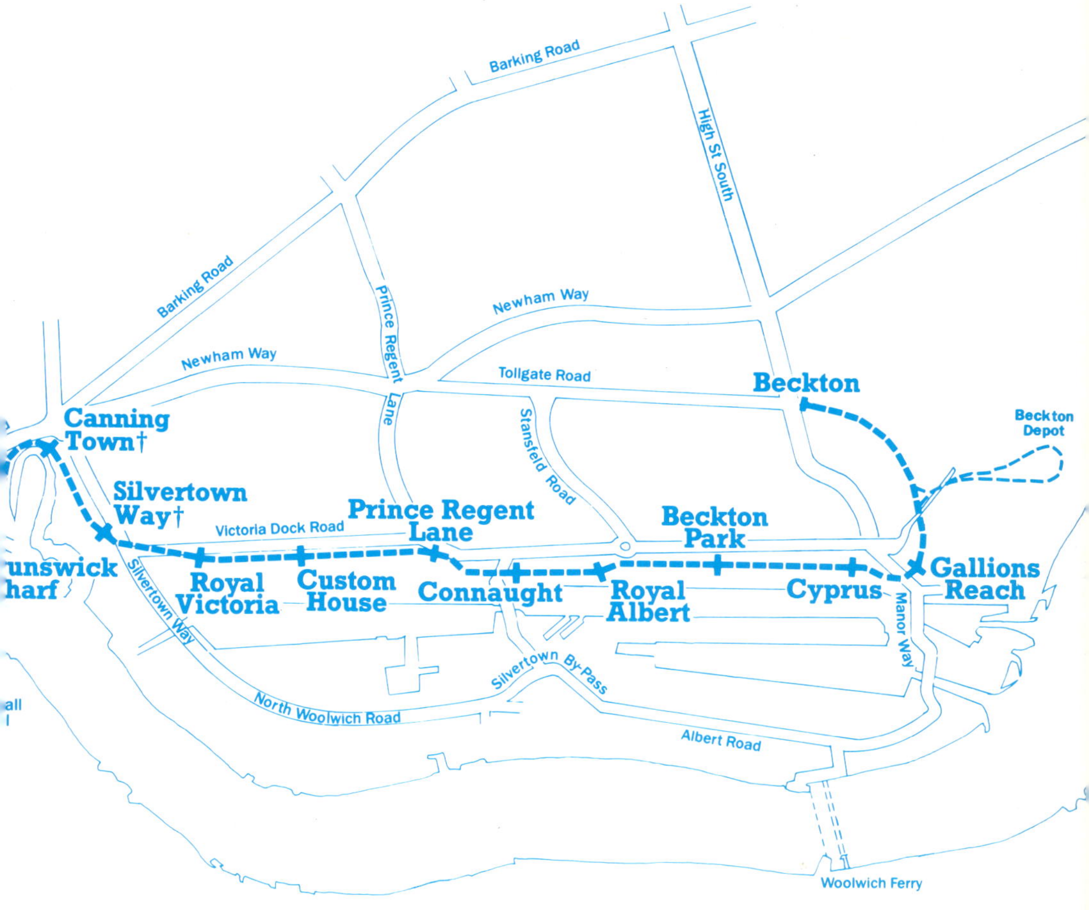

With the Bank and the Beckton extensions open the current plans are that journeys will operate: Bank and Crossharbour; Bank and Beckton; Stratford and Crossharbour; Stratford and Island Gardens. It is likely that Tower Gateway will be retained as an emergency alternative City terminus and also for use, principally at weekends, as a visible tourist link to Docklands and Greenwich for which a through service to Island Gardens would be operated.

STATIONS

The stations designed and constructed for the Initial Railway look more like mini-mainline or heavy suburban service railway stations rather than typical minimal Light Rail stops.

Yet they are still relatively cheap to construct and there are lots of them — 15 on 7½ miles of route. Although designed around a standard kit of modular parts these parts are re-arranged to suit individual sites.

The station on Heron Quays featured as an early typical station both in perspective drawings and in model form. In fact, only three of the 15 Initial Stations are like Heron Quays. These are West India Quay, South Quay and Mudchute. Heron Quays features an all-new station on new elevated trackwork with a small concourse where two staircases, two lift tower bases and two ticket machines are located. All of the remainder are different though three stations on the northern route have strong similarities, one with another.

KEY STATION DESIGN FEATURES

Tower Gateway A one-off that is quite unlike any of the other stations. L-shaped in plan and elevated, it is approached via a glass rotunda at the foot of two escalators with flanking stairs; the concourse is end-on to an island platform with two tracks, one each side. Two hydraulic lifts are located opposite the two ticket machines across the width of the concourse itself.

Shadwell This station is unusual in being an island platform design squeezed on to the existing viaduct where two British Rail tracks and two DLR tracks are accommodated. It has a single lift at the extreme western end which has its lower door opening located near Watney Street.

Limehouse is fashioned around and on top of the 1840 era viaduct beside British Rail Limehouse station.

Westferry is unusual in that the platforms are over Limehouse Causeway to accommodate access from either side of the street, so there are two separate concourses underneath, each with a lift, ticket machine, validator and staircase.

West India Quay The Heron Quays pattern: elevated, all new, with the concourse extending southwards to the quayside. The final emphasis on the direction of approach of passengers to this station is not yet known.

Heron Quays is the role model for the others that are similar.

South Quay Similar to Heron Quays.

Crossharbour Although similar to Heron Quays, the local ground level is much higher where the concourse is located, with separate steps and ramp access down to the roadside.

Mudchute Similar to Heron Quays.

Island Gardens Another one-off, featuring an elevated two-platform V-shaped layout. The reinforced concrete structure was finished with false brick arches using old stock bricks and, although not designed to provide shop units, one arch was extensively converted into a DLR exhibition/shop unit in July 1987.

Poplar is also a one-off, featuring three tracks serving two island platforms with two western-end located lifts similar to Shadwell, but the station is built upon an earth-filled, wall-retained structure rising up from ground level at the eastern end.

All Saints An overhead, roadside concourse feeds down, not up, to platforms via separate staircases and lifts to each side platform. The east side platform and lift is slightly further south than the west side one.

Devons Road is very similar in layout to All Saints, but without the displaced platform location.

Bow Church is similar to Devons Road.

Stratford DLR trains terminate in the previously unused bay platform 4.

Each station is a kit of parts with standard platform sections, canopies, stairs and lifts. When Heron Quays becomes encased in a new development the station components can be unbolted and used to provide additional features at other stations.

Because of the limited space available, an island platform was built at Shadwell, with the up line cantilevered off the old brick viaduct and the platform constructed on the space released. The CCTV monitoring the lifts can be seen at the end of the canopy whilst the platform CCTV is on the pole at the other end of the platform.

Poplar station is currently the largest on DLR and was built as the interchange station for the proposed Tower Gateway to Beckton and Stratford to Island Gardens services. With expansion in mind the level section of elevated track stretches some 30m beyond the present platforms, making lengthening a simpler process.

All station platforms have to be built on a straight section of track to keep to a minimum the gap between the train and the platform, ensuring easy access for pushchairs, wheelchairs and shopping trolleys. A simple hydraulic lift serves each platform.

Beneath the station canopies are simple perch seats, train describer displays and passenger alarms. The latter are linked direct to the Control Room and trigger the recording of CCTV pictures of that platform as well as sounding an alarm buzzer on the control desk. Use of glass and polycarbonate platform backs gives waiting passengers a superb view over surrounding areas.

Cutting stations like All Saints contain the same kit of components as elevated stations. The blue drum structure is the Telecoms Room containing signalling and communications equipment. The tranquillity of this scene is in complete contrast to the busy East India Dock Road above the station!

STATION FACILITIES

Initial Railway stations feature 30-metre (98½ft) long platforms with large and distinctive curved 9-metre (30ft) long canopies; staircase access for the fit and able; hydraulic lifts, one to each side platform and one to each island site; perch seats; 'Next Train' indicators with a 24-hour clock; press-to-speak passenger alarms in each lift and on each platform; a high level of artificial illumination; two self-service ticket machines and two validator columns; closed circuit TV cameras to monitor each platform.

Each station can, and probably will be, enhanced with a package of further modular units such as 3-metre (10ft) canopy segments, covers to staircases, extended platforms and also with escalators. Each station is designed to have escalators added (except Tower Gateway which already has two) and these will be built onto the outside of the existing stair structures in most cases. Voids to house future escalators within existing infrastructure already exist at Shadwell and Poplar, where station locations prohibit outside-access escalators.

A raft over the railway at Devons Road holds the ticket issuing machines in the centre and the two validation columns, one by each lift and stair entrance to the up and down platforms. With unstaffed stations eye-catching explanatory signs must tell first-time users and remind regular passengers of services offered, ticketing procedures and access to platforms. Prominent coloured enamel signs face the traveller and a tall illuminated 'Docklands' sign emphasises the station's location to passing traffic. The high level of lighting and the open layout makes for a bright secure station.

SIGNS

The contract signed with GEC-Mowlem to design and build the whole railway covered the provision of passenger signs for a safe railway. In the event, the client, DLR, was not satisfied with the overall number of signs provided nor in many instances with the quality of fixtures and fittings. A signing survey was carried out in April 1987 and a re-signing programme commenced in June which unfortunately was not complete when the railway opened. The main points of the re-signing programme were to provide for a better quality of graphic presentation; more signs; more comprehensive signing; and signs that would be likely to last longer at exposed unstaffed stations where vandalism might prove to be a problem. In practice vandalism has been minimal.

The programme has based sign design on the 50mm (2-inch) module adopted by LRT for London Underground's signing. However the Docklands' Rockwell typeface has been used rather than the Underground typeface. Signs are stove-enamelled, not vitrious-enamelled which is much more expensive, because within two years many signs will have to be replaced to incorporate details of the western and then eastern extensions. The hardest signing to design was that required to go next to the Thorn-EMI ticket machines. Initially it was decided to use English language verbal sentence instructions and to cover every single topic. Subsequent experiments at Island Gardens in October 1987 saw pictorial instructions used. Unfortunately, at many locations on the railway there is a lack of other structures on which to hang or erect signs.

The impressive entrance of Tower Gateway station with its stair and escalator shaft leading up from the main dome emphasise its importance as the link from the City of London to Docklands. Behind the station are the station approach tracks to BR Fenchurch Street.

Although Tower Gateway station is the only station currently equipped with escalators, the construction of all station sites allows for escalators to be added, normally adjacent to stairs, at a later stage if funds permit.

Island Gardens station is built on a short wide site between two roads, hence the splayed formation of the two side platforms. As the other 'gateway' station design its glass and polycarbonate dome match the red brick and glass dome of the nearby foot tunnel under the River Thames to Greenwich.

The centre of the Island Gardens dome contains the lift whilst the stairs spiral around the edge, both leading up to the connecting walkway to the platforms. It was here on 30 July 1987 that Her Majesty The Queen opened the DLR.

CITY EXTENSION

The success of the Isle of Dogs as a development area, due at least in part to the coming of the DLR, has encouraged many big office and commercial projects. The largest scheme so far is on Canary Wharf and the adjacent land to the west as far as the river bank. Based around an international financial trading centre with hotels, shops and other office developments, it is planned on a huge scale. The original key statistics when the scheme was launched were three ultra-high tower blocks, 10 million sq ft of office space and up to 40,000 jobs!

For this to succeed the developers, originally a consortium of three investment banks, but now a Canadian development company, wanted a direct line to the heart of the existing City of London using the DLR.

This is being achieved by a 1.5km extension to the railway running westbound to a new station under King William Street in the City, being built under the London Docklands Railway (City Extension) Act 1986 which received Royal Assent in December 1986.

This will give direct links to the Underground stations at Bank and at Monument to connect with the Central and Northern lines at Bank and with the Circle and District lines at Monument. The British Rail Network SouthEast Waterloo and City line will also be connected via Bank station. When the extension opens at the end of 1990 (a single track and platform; the second track and platform opens in 1991), Canary Wharf will be only 9 minutes by DLR train from the heart of the City.

The Bank extension leaves the initial elevated railway at a point near Royal Mint Street. The tracks will plunge underground down a six per cent gradient into a cut and cover tunnel, then in deep level bored tube to the Bank. Typically, the extension will be built at a depth of between 35m and 38m and 1.2km of the 1.5km extension is underground. Tunnels will be built to a 5m diameter — sufficiently large to take existing DLR stock and other typical Light Rail vehicles. The Contractor for the tunnelled extension is Edmund Nuttal, Camberley, Surrey.

A working site at Royal Mint Street has been retained from the days of the initial railway construction. The tunnelling drive will begin from this site. The end of the contract will be in the autumn of 1990 with a small maintenance facility kept here for a further two years. Several working sites have been or will be set up along the route of the extension: these are grouped towards the western end of the new line in the Monument and Bank station areas.

At the Fish Street Hill site a temporary access shaft is being sunk to allow work to start on the concourses and passages for the DLR Bank station. This access will later on allow the tunnelling shield to be enlarged at the end of its drive from Royal Mint Street before continuing on with the work for the larger station tunnels. A separate permanent ventilation shaft will also be sited here. This site will be occupied until the autumn of 1990.

Of the other three main sites, Bucklersbury near Queen Victoria Street will contain a temporary shaft to give access to an existing tunnel at the end of British Rail's Waterloo and City line. From this existing tunnel will be built all of the passenger tunnels, escalator tunnels and lift shafts necessary to link the Waterloo and City line, Northern and Central lines. In Lombard Street an extension to the existing London Underground ticket hall will be built. There will be a new entrance stair from the street and an escalator from the ticket hall which will be linked by new tunnels to each of the lines at Bank station. In Lothbury there will be a permanent ventilation station for the DLR; the shaft will allow fresh air to be drawn down to a fan chamber at the foot of the shaft.

The DLR platform will be at a very deep level below the existing Northern line platforms but lift access will be provided from street level in order to preserve the existing standards of disabled access.

Docklands

Docklands Light Railway
City Extension Tunnels and Bank Station

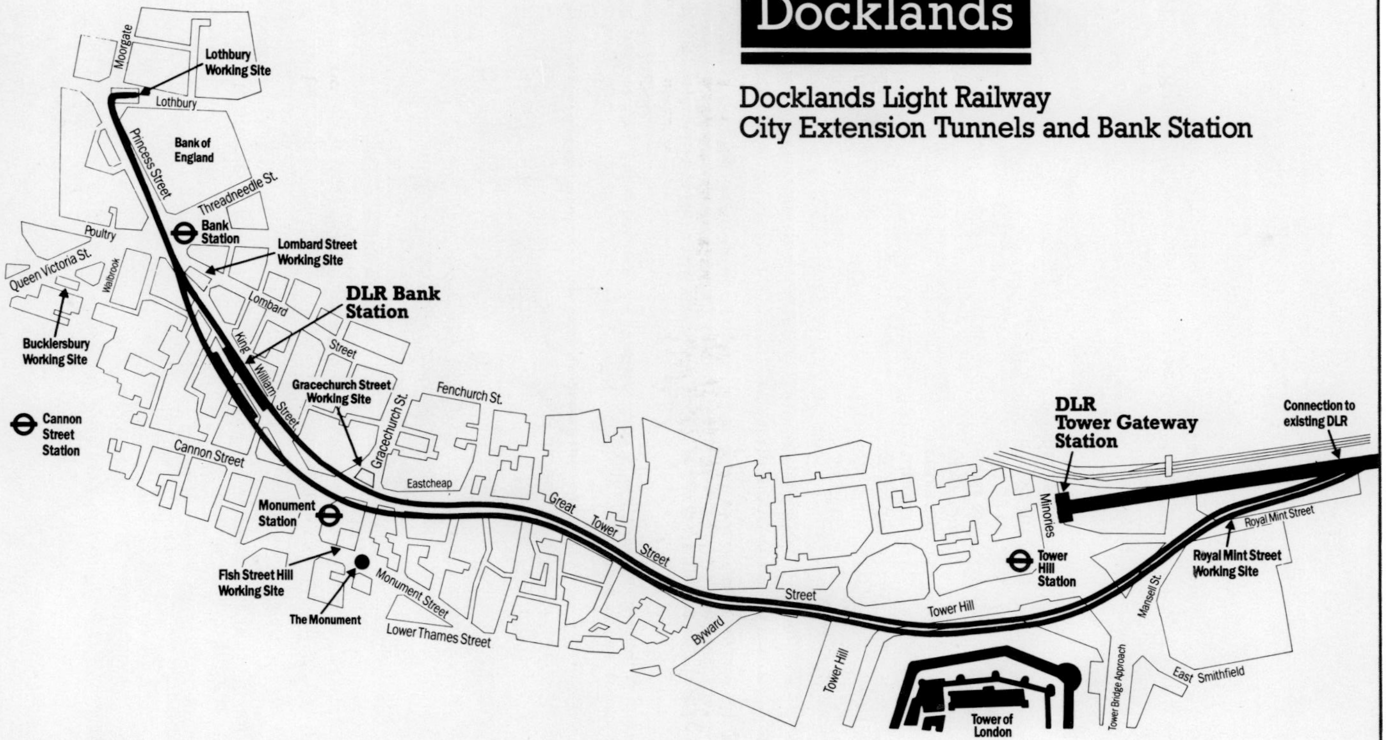

Published by Docklands Light Railway Limited,
P.O. Box 154, Poplar, E14 9QA
Telephone: 01-538 0311

Map no. 12 November 1987

The large diameter of the station and running tunnels is emphasised in this artist's impression of DLR Bank station. Lying beneath the Northern Line it will have the deepest platforms below sea level in London.

With the construction of Bank DLR station the access and interchange routes within Bank/Monument stations will also be improved. A long overrun tunnel from the DLR platforms allows a failed train to be moved from the area without disrupting traffic and a new lift shaft ensures disabled access.

INITIAL RAILWAY UPGRADING

As well as the dramatic Bank extension tunnelling work, substantial changes and improvements to the capacity and capability of the initial, open-air railway are also underway. A contract to handle this work has been placed with GEC-Mowlem. Principal features include doubling the platform lengths at Shadwell, Limehouse, Westferry, West India Quay, Heron Quays, South Quay, Crossharbour, Poplar, All Saints, Devons Road and Bow Church; Canary Wharf will become a major station with escalator access to the adjoining developments — the station built for the Initial Railway on this site was demolished at the time of opening the railway; reversing sidings will be built south of Crossharbour station near to Eastferry Road; ten, then a further ten, trains will be built; other structures will be strengthened as required and changes will be made to the railway signalling.

BECKTON EXTENSION

Throughout the development of the DLR there has always been an aspiration to continue downstream to the Royal Docks, an area of sprawling water frontages and space much greater than in the Isle of Dogs. It was for this extension that the north curve of the North Quay delta junction was included, and, with the anticipation that traffic levels would justify two-car trains, City branch platforms were designed for easy lengthening.

With the increase of confidence and investment in the Isle of Dogs, much grander schemes began to leave the drawing board for the Royal Docks as well. The London City Airport opened in November 1987 and together with extensive housing development in the Beckton area forms some of the early infrastructure attracting new development.

In November 1986 a Bill was deposited in Parliament by LRT for the construction of the extension. This has been subject to some delay — not least by the General Election in 1987 and the rejection, in March 1988, of the southerly (Leamouth) alignment by the Commons Select Committee. Royal assent is anticipated by the summer of 1989, and would incorporate the deviation through Canning Town necessitated by the Select Committee's decision.

The extension is likely to have 13 stations, of which most will be simple structures similar in concept to those already built, but with some refinements arising from operational experience. One or two of the stations may be incorporated into new developments, and interchange with the BR North London Link — itself facing upgrading — is likely to be at new stations at Royal Victoria and Prince Regent Lane. A minibus shuttle from the latter point will also carry passengers to the London City Airport terminal.

Because the trains needed for the upgraded initial railway will fill the Poplar OMC site, a new depot will be built near Beckton to service this extension and further foreseeable expansion. This will become the major maintenance depot, leaving Poplar to handle inspections and casualty work. The site chosen was once part of Beckton Gasworks.

Services on the line will be of two-car trains during the day (and possibly three cars ultimately) at peak frequencies of up to four minutes: some trains may however turn back at Connaught. Because of the higher frequencies on this and the other branches the existing flat delta junction at North Quay will be inadequate and will be replaced by a spectacular flying junction which, with track remodelling through Poplar, will result in the complete segregation of City – Beckton trains from Stratford – Isle of Dogs services. This will also afford an excellent cross-platform interchange between the lines at Poplar, giving good connections between Beckton and the Isle of Dogs and between Stratford and the City.

North Quay Junction may well be an extreme example, but it will not be the only case of spectacular civil engineering: long spans will be needed to cross the River Lea and at several places the railway will be constructed along with new road developments, including the approaches to the proposed East London River Crossing, carrying a major new highway soaring above the Thames.

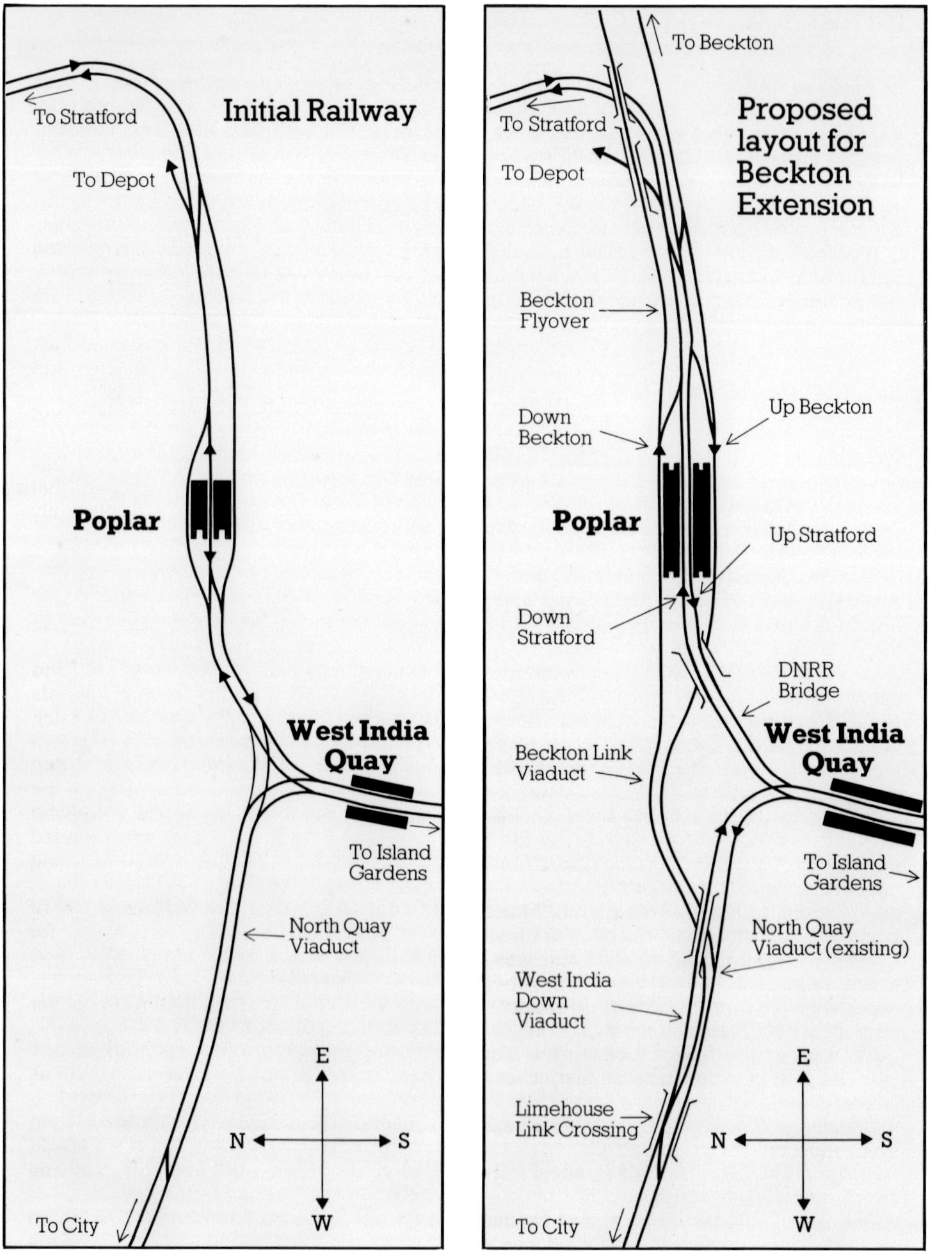

Track Remodelling North Quay Junction and Poplar

OTHER EXTENSIONS

In addition to the main City/Bank and Beckton extensions, reasoned proposals for the DLR to be extended to various other locations have been proposed and are under discussion. These include:

1. Beckton to Barking (via one of two routes).
2. Custom House (on the Beckton extension) to London City Airport (also via one of two routes).
3. Custom House to North Woolwich, Woolwich and Thamesmead, south of the river.
4. From the Beckton extension near the proposed Cyprus station via the East London River Crossing to Thamesmead.
5. From Stratford to North Woolwich, partly in parallel and partly by taking over the British Rail track.
6. From Thamesmead due westwards towards to the south end of the Blackwall tunnel, thence via tunnel to the Isle of Dogs at Crossharbour, on an east-west axis, and westwards again under the river towards Rotherhithe or the New Cross area.
7. From the Isle of Dogs southwards under the River Thames, thence to Greenwich and towards Lewisham British Rail station.

This last plan is actively supported by the London Borough of Lewisham which has publicised the extension considerably during the second half of 1987, to the extent that it is now regarded as 'fact' by many local residents! LRT and the LDDC are currently reviewing the plan.

INITIAL OPERATING EXPERIENCE

The acid test for any railway is to attempt to carry passengers — fare-paying ones who can grumble! Many people visited the railway during the hectic building period and towards the end of its construction visitors were selectively taken on test train trips. Many such visits were actually designed around specific incident tests and trials. For example, a Chartered Institute of Transport visit which joined the railway at Stratford found itself being de-trained on Limehouse viaduct twenty minutes later! A comprehensive programme involving the emergency services in incident training also took place during the spring and summer of 1987. Combined police, fire brigade and ambulance exercises were carried out at a number of locations at Bow Curve, Limehouse viaduct and on the Docks Crossing bridges for example.

The first public day of operation, Monday 31st August 1987, saw the severest test of the system with upwards of 40,000 people trying out the railway. The daily target for operation at the end of the first year was 20,000 users per day!

In summary, initial operating and general experience has revealed:

1. The Additional Fares and on-train ticket checking policy works well and much better than many critics expected.
2. The lack of a through Stratford — Tower Gateway service is commented upon.
3. The station concourse-located ticket machines are not reliable and not regarded as user-friendly.
4. Trespass, vandalism and general passenger misbehaviour is much less of a problem than expected by some.
5. The door design, as expected, is not liked by passengers though incidents involving the doors are few.
6. Station hydraulic lifts are used by the fit and able, not just those actually needing to use them: the disabled, wheelchair users, shopping-laden and people with young children. This causes some overloading and subsequent failures.
7. Tourist, or rather visitor, traffic at weekends is much heavier than expected with standing loads on Sunday afternoons being the norm not the exception! Much more tourist traffic originates at the Greenwich end of the railway than was expected.

STAFFING

From the outset it was clear that the staffing regime of DLR would be different — different from the Underground and indeed different from any transport operation in Britain.

Here was a golden opportunity: a brand new railway with an ability to have modern staffing conditions appropriate to the closing decades of the twentieth century (and, hopefully, well into the next). Yet it was a very small system with its own problems of the diseconomies of small scale. DLR had all the risks of a big railway, but how could it afford to have on duty round the clock, for instance, a rolling stock technician, a signalling technician, a buildings person, some track staff, an electrician, an electronics expert . . .? The list would be endless, with all of these people waiting for something to go wrong.

Very early in the planning stages of the railway some principles became clear:

1 Staff would have to be flexible — no rigid demarcation lines, no great rifts between different departments and specialisation only where required by the needs of the job.
2 People working on the railway should live locally as far as possible. There are always problems when shift workers have to travel long distances but with DLR there would be the added need for a small local railway to be run by local people.
3 Salaried status — there are some enormous benefits all round from abandoning the Victorian concept of hourly paid workers. There was a very strong desire to avoid a 'them and us' situation developing.
4 Staff representation should be straightforward, effective and appropriate to the size and philosophy of the railway.

These early desires translated into a package of conditions which many people clearly found attractive, with most jobs heavily oversubscribed — especially in the case of traffic staff where over 2200 applicants were chasing the initial 54 jobs. All were salaried, over half from the immediate Docklands area (with most of the remainder from the rest of East London) and most recruited to fill jobs with varied and flexible job descriptions. Staff representation was through elected councillors and although many retained individual union membership from previous jobs or joined unions, collectively the workforce opted against union representation at negotiating level in a ballot held just before public opening.

The effectiveness of every single member of staff is crucial to the future of DLR, but from the passenger viewpoint the Train Captains are in the front line and it is worth looking a little more deeply into this group of people.

Train Captains are actually known in DLR as Traffic Assistants. Why? Well, although the vast majority of their time is spent on the trains their brief is in fact much broader. They can, for instance, mount ticket checks other than on trains, they can become involved in cleaning, in train fault finding and inspection, in attending schools and exhibitions to represent DLR and they can become involved in administrative work. Flexibility is enhanced by the fact that Traffic Supervisors, Control Room Assistants (who are the people that keep passengers and the media informed of how DLR is running) and some management people are also qualified to operate as Train Captains.

Many DLR passengers — and neighbours — once viewed automation with suspicion. Would there just be robots? No — DLR was determined to take the latest in automation to actually humanise the railway. Hence a member of staff, freed from the need to drive and able to meet the customer, to check tickets, to lend a helping hand where needed or just have a chat.

Clearly, DLR did not need 'train drivers' for this job. Although the people recruited needed to become fully conversant with railway practice, safety rules and driving techniques, the overriding need was for staff who enjoyed working with their customers, people with personality. This could not be taught — so recruitment was aimed at people from shops, restaurants and so on. The background of the Captains in the event was remarkably varied with three ex-bus conductors the only ones with any transport background. The Traffic Supervisors reflect the need for greater transport experience and half of the twelve originally recruited had this background, with three from British Rail and three from London Buses.

From all angles this policy has been an outstanding success, with passenger complaint almost unheard of, against a continuing flood of commendations. Passengers are kept happy — even on the odd occasion when services have hiccupped — tourists have been given a commentary of the passing and rapidly changing scene, and vandals have turned their unwanted attention away from the railway.

Maintenance staff, understandably, have their roots rather more firmly in railways, with many having come from British Rail or the Underground. Again, they have adapted to the flexibility required and excitement gained from working in a small but highly accountable organisation. Between them the small teams keep the buildings, tracks and trains fit for the eyes and feelings of nearly 20,000 passengers each day, and under the eye of what so often seems the nation's — and the world's — press.

Finally there are a few jobs, such as cleaning, escalator and lift maintenance and similar specialist tasks, which are most effectively left to contractors. For the most part though, DLR's services are essentially home made in Poplar.

A Train Captain's responsibilities extend beyond the train to include checking the station platforms — a job made easier by the open platform layout. Closing the train doors can be done from any door position on the platform side, allowing the Train Captains to move about in the train as they wish.

MARKETING

Although it has been suggested that the railway will market itself this is not, unfortunately, the case!

Whilst it is true that the incoming business community will grow larger and larger, bringing more and more commuters, these will be essentially peak hour travelling groups, so that as on other railways they will stretch resources to the maximum. There remains a great need to fill train seats — and floor space — at other times of the day. The current marketing plan recognised four main marketing segments to aim at: commuters, shoppers, tourist and leisure use, and disabled use. Obviously, there are many other groups making up the likely total markets.

Much of the pre-opening marketing activity has been involved with informal contacts with businesses coming to the area. The railway has been helped enormously by the all-powerful effect of promotion of Docklands by the London Docklands Development Corporation. Without this vast effort it would have been impossible to promote this little railway without a lot more money being spent on marketing — and this has not been available in this cash-limited project. There has been since the start of the project an indication of potential balance in the traffic: off-peak shopping traffic to Stratford, an important regional shopping centre; tourist traffic from the Tower of London to Greenwich, and vice versa, via the railway; some growing evening leisure use with the increasing number of facilities to be found in the Isle of Dogs and Wapping areas, as well as the obvious and always present growth in commuter traffic into and out of Docklands itself.

There are clear signs already that the railway is seen as an integral part of the Docklands community and not as something just grafted on.

PUBLIC RELATIONS

In order to be successful in an area of great and rapid change, the DLR has planned to be accepted by the widest spread of local population for the benefits that it will bring to their lives. Without at least a measure of support, the novel package of technology and operating style used on the railway could have been put at risk and less likely to work efficiently. It was recognised that there were many risks involved in not having local support, including the effects of having a degraded passenger environment to contend with. Abuse of the railway infrastructure could easily have led to yet more abuse. The DLR project team from the earliest possible date chose to be based in Poplar just like a local business; this was achieved in December 1984 within weeks of the contract to build the railway being signed.

Contacts were set up at an early date with the local police, fire brigade and ambulance services. There is a continuing dialogue with the local crime prevention officer as well as the extensive contacts with the British Transport Police. Local planning group meetings take place with all local authorities with powers to reflect the local environment. Letters, meetings and discussions take place with the local action groups and social groups such as the Docklands Forum and the Association of Island Communities as well as more formal meetings on an ad hoc basis with local and prospective local businesses. In the very tight construction timescale of the railway's development over 5,000 people visited the railway before it officially opened to the public.

In 1984 it was recognised that the railway needed its own local voice in the area and this led to the publication of 'DLR News' as a low-cost tabloid newspaper delivered door-to-door. This newspaper has always been well received; each edition produced a large amount of correspondence and fresh contacts.

Left **Service information leaflets provide details of fares, frequencies and first and last trains.**

Below **An attractive A4 size booklet published by DLR at £3, giving details of places of interest on or near the railway route.**

Docklands Light Railway News

WIN A TENNER – Back Page

NUMBER 6 JULY 1987

60,000 COPIES DISTRIBUTED THROUGHOUT THE AREA

FARES FAIR!

Getting your ticket to ride in two simple stages

One, buy a ticket. Two, validate it.

BUYING a single ticket to travel on Docklands Light Railway is a two-stage process.

One — buy a ticket. Two — when you want to travel (you have the choice, it can be minutes, hours or days later) validate it!

That is, stick it in a cashcard-style blue totem pole machine with an electronic brain. It will return you the ticket dated and time-stamped for travel in a flash.

Officially we call the totem poles validators, but passengers will feel the simple process is more like "checking in" for travel as and where they want to go somewhere in or beyond Docklands.

Ours to be just the same as on the Underground

THE FARES to be charged on the Docklands Light Railway will be "the same as the Underground and fit into the London zonal fares pattern," Managing Director Cliff Bonnett announced this month.

Docklands stations will fit precisely into Zones 1, 2 or 3a of the familiar fares pattern of the Underground. Nearly all of the initial railway is in Zone 2 and most local trips will cost just 40p for an adult single ticket.

This announcement has scotched rumours of "high fares" or "premium business fares" which have been circulating in the area — especially on the Isle of Dogs — for some time.

Inside, we explain the full fares facts with examples of what many trips will cost.

Meanwhile, as our picture shows, we're getting ready for the Off. Test trains are now running a frequent ghost service on all parts of the railway. Car 06 is pictured at All Saints Chrisp Street Market station by the East India Dock Road.

Left **The issue of the DLR newspaper published just prior to opening.**

The company has an extensive photographic library of views of Docklands taken during the construction of the railway. Serious historical researchers and people wishing to purchase photographs should write to Docklands Light Railway Limited, Public Relations Department, PO Box 154, Poplar, London E14 9QA to make an appointment. There is an administrative charge of £8 minimum per visit and copies of black and white prints, colour slides or colour prints are charged at cost.

THINGS THAT MIGHT HAVE BEEN

Many passengers and visitors have been intrigued by earlier references to pictures or statements about the DLR which have not come about. These might-have-beens include various train designs, various routes, station sites and names, liveries and even the name of the railway business itself!

The industrial design of the trains was first established in 1983 when design consultants made careful proposals. However, the change in 1984 to the single-package tendering process meant that the commercial bidders had to make proposals based on what was available from the world Light Rail industry or could be developed in time and within budget, to meet the Performance Specifications laid down by the clients. This specification related more to capacity and performance objectives and did not otherwise dictate overall vehicle appearance. Some bids were based on derivatives of West German LRV, others such as the French and Canadian proposals featured smaller individual vehicles, very similar to those operating in Lille and Vancouver.

The most significant might-have-beens concerning routes include the two plans to run the railway towards Mile End Underground station. It was first intended to have a street-based line running from the Limehouse area to Mile End along Burdett Road. Later the Poplar to Bow railway line, as now used as part of the DLR, was intended to carry the railway northwards for the start of the nothern route and then the line was to have turned southwest at Bow along the Mile End Road itself to Mile End Underground station. The ultra-tight timescale and large number of authorities and utilities with interest in sub-surface London streets worked against both possibilities. There is not much doubt that Stratford, a railhead for much of southeast Essex, as well as being a shopping centre and bus interchange, is a much more worthwhile link for the Docklands system.

A number of station sites have changed during the brief few years the railway has been under development and many station names have changed as well! It is often forgotten that the total number of stops planned has been increased, aided by the competitive tendering process in 1984. The earlier GLC maps published in its newsletter are markedly different from the current DLR. There are more stations now and changes of route, and more than half of the current fifteen station names have been changed as well. The DLR team retained a preference for local geographical names rather than area or district names with additions — north, south etc — which are a feature of the larger scale and less intimate Underground railway network. Of the current fifteen names, the origins of most of them are obvious and can be discovered from local street maps or a London A-Z atlas. Some do need further explanation, though. Mudchute station, previously Millwall Park, is named after the adjacent high level plateau now used for City Farm and allotments as an open green space. The site itself was originally used for silt taken from the Millwall dock. The recent change of station name from Millwall Park to Mudchute followed local representations expressing concern that visiting football club supporters looking for Millwall's football ground would alight at Millwall Park station only to find that Millwall football club moved south of the river some years ago; a discovery that would likely lead to irritation and damage to local property!

Poplar station, earlier referred to as Interchange, was designed to handle the east-west access Beckton to City trains as well as the initial railway's Stratford to Isle of Dogs service. With a Beckton service in place the earlier name would have some of its original meaning restored and with the coming of this extension Poplar will probably be renamed Poplar Interchange.

Regarding liveries and railway names, everyone has an opinion! The original design proposals made by consultants featured white-liveried units. This livery was not liked by the embryonic operating team for the expected difficulties in keeping the paintwork clean. After GEC-Mowlem won the contract to build the railway its designers produced several distinct livery choices for the Joint Board to approve and eventually a blue, red and white scheme was adopted. Before this happened, GEC-Mowlem had three main proposals to offer and these can be summarised as the blue, red and white design ultimately adopted, a yellow and grey scheme, and a triple blue livery with traditional railway industry orange

During the first half of 1984 some pre-emptive bids to construct the railway were made by British-led groups, in advance of any invitation to tender. One such proposal was put forward by Metro-Cammell and Westinghouse as 'METRO-DART' (Docklands Area Rapid Transit); these views show alternative concepts for the appearance of the rolling stock. The adoption of a third-rail system came late in the period of negotiation with short-listed tenderers, at the behest of the LDDC which did not wish to see overhead ironmongery spoiling dockland views.

Later in 1984 METRO-DART became one of the five consortia bidding for the construction of the railway. Note by this time the adoption of the barrel-vaulted station canopies which were to become a feature of the initial railway.

These photographs are produced by kind permission of Metro-Cammell

This view shows an impression of the attractive vehicle offered by METRO-DART during the late pre-contract period. Note the abandonment by this time of the overhead power supply and the elegant bridge structure depicted against the backcloth of Canary Wharf as it was in 1984. It should be pointed out that this and all designs shown in this chapter reflect contemporary thinking at that time and are no longer current.

lining. As first presented the successful scheme featured a stylised arrow and circle motif; the yellow and grey scheme incorporated a red 'Clipper Rail' motif intertwined with a map of the River Thames curling around the Isle of Dogs. The triple blue livery included a nautical chainlink device surrounded by a double ended arrow. Even then the logotype and graphics treatment was part of a separate series of competitive work implemented late in 1986 and based around the logotype 'Docklands' rather than any contrived symbol or squiggle, distinctive or otherwise.

In fact, the blue, red and white livery was originally designed and proposed to the DLR project team by Vicky Ferguson, a design student at Middlesex Polytechnic. Following a dialogue between the GEC-Mowlem designers and the student, this scheme was adopted as a proposal to be offered to the DLR clients. The logotype 'Docklands' — and all other large scale typography — is set in Rockwell Medium as proposed by Cooper Thirkell Associates of Putney, who made this proposal as part of their bid to win the railway graphic design work competition held in 1986. Subsequently, these consultants were not retained by the railway, but another practice, Burton, Soar, Chambers, who had designed some of the earlier DLR maps, were employed to implement this work, again using the Rockwell typeface.

The railway we have today is the end product of much thought and talent. The chosen designs have been well received and give the DLR a distinctive identity which yet blends in well with the Docklands development as a whole.

Design consultants prepared a full set of proposals for the graphics in 1983 featuring distinctive triangular sail-shaped signs on which London Transport's Johnston typeface was used. Illustrated is a station sign for Heron Wharf (now Heron Quays) and an associated route map for use at the same station. Of the stations shown, three have not been built, Limehouse (east of the present station of this name), Fawe Street and Pudding Mill Lane, though the latter two are protected sites where stations can still be built when and if finance becomes available. A site at Carmen Street has now replaced Fawe Street. Station name changes include Tower Hill, now Tower Gateway, Stepney East, now Limehouse, West Ferry Road, now simply Westferry, North Quay, now West India Quay, Heron Wharf, now Heron Quays, Glengall Grove, now Crossharbour, and Millwall Park, now Mudchute. Poplar station is an additional site next to the Operations and Maintenance Centre; East India Dock Road station is now All Saints, Devons Street should have read Devons Road, and Bow Road is now Bow Church.

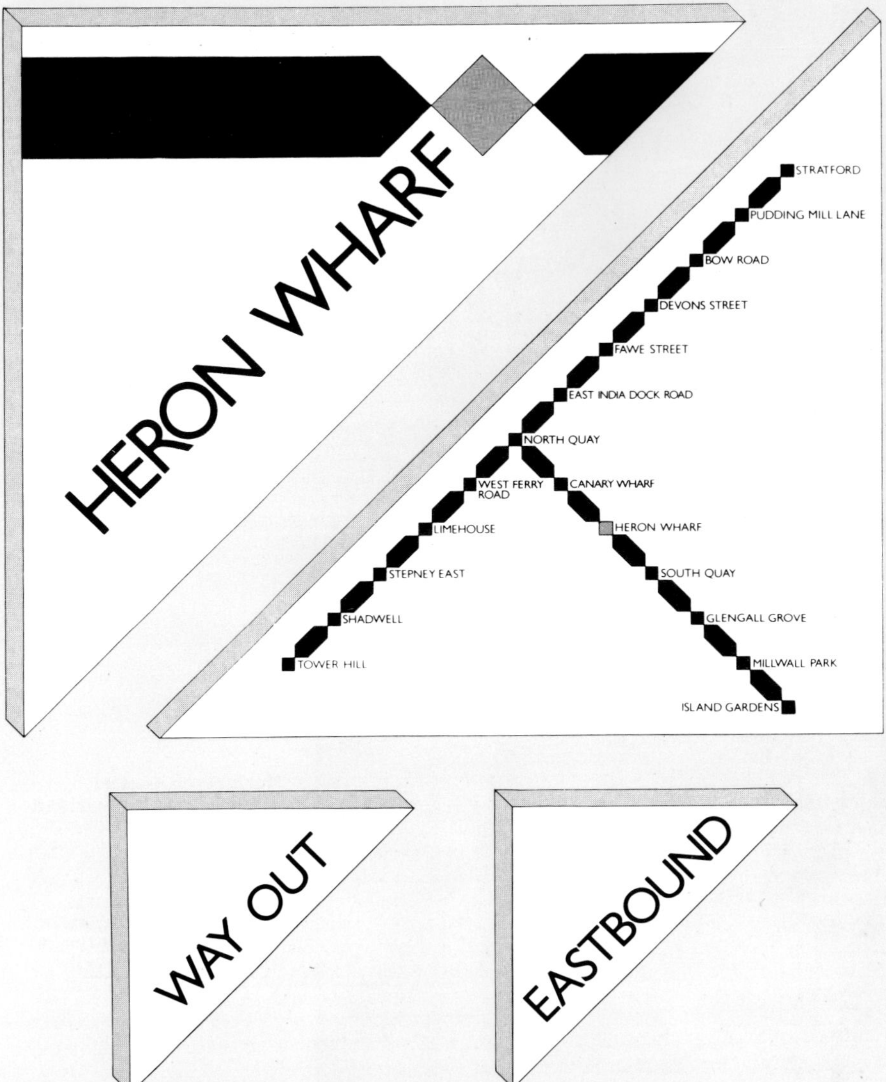

A design competition with students from Middlesex Polytechnic produced some interesting 'might have beens' as well as some actual ideas incorporated in the final railway. This design for Westferry shows an enhanced station with escalators and extended canopies.

Although the DLR makes extensive use of painted steel, glass and plastic, alternatives of wood and brick were put forward, offering a more traditional station image.

This is the origin of the present DLR train livery, part of an overall design concept produced by Vicky Ferguson. Highlighting the vehicle doors with a bright colour is particularly helpful to the unfamiliar passenger, whilst the strong blue and red make the train very visible in the city landscape.